155.67 C834b

The course of later life:
research and reflections

DISCARD

The Course of Later Life
Research and Reflections

Vern L. Bengtson, Ph.D., is Director of the Andrus Center's Gerontology Research Institute and Professor of Sociology and Gerontology at the University of Southern California. He received his B.A. from North Park College and his M.A. and Ph.D. from the University of Chicago. Dr. Bengtson directs a pioneering Longitudinal Study of Three-Generation Families at U.S.C. as well as continuing research on the sociology of the life course, socialization, ethnicity, and aging. His publications include *The Social Psychology of Aging, Youth, Generations, and Social Change* (with Robert Laufer), *Emergent Theories of Aging* (with James Birren), *Grandparenthood* (with Joan Robertson), and *The Measurement of Intergeneration Relations* (with David Mangen and Pierre Landry). Past Chair of the American Sociological Association's Section on Aging, Dr. Bengtson is President-elect of the Gerontological Society of America. He has twice won the Reuben Hill Award for outstanding research and theory on the family presented by the National Council on Family Relations. He most recently was awarded a MERIT award from the National Institute on Aging for his longitudinal study of families.

K. Warner Schaie, Ph.D., is the Evan Pugh Professor of Human Development and Psychology and Director of the Gerontology Center at Pennsylvania State University. He received his B.A. from the University of California at Berkeley and his M.S. and Ph.D. from the University of Washington. Dr. Schaie is the author or editor of a dozen books and over 150 journal articles and chapters related to the study of human aging. He is the recipient of the Distinguished Contributions Award from the Division of Adulthood and Aging of the American Psychological Association, and of the Robert Kleemeir Award for Distinguished Research.

The Course of Later Life

Research and Reflections

Vern L. Bengtson, Ph.D.
K. Warner Schaie, Ph.D.

Editors

Springer Publishing Company
New York

Copyright © 1989 by Springer Publishing Company, Inc.

All rights reserved

No part of this publication may be reproduced, stored in a retrieval system, or transmitted in any form or by any means, electronic, mechanical, photocopying, recording, or otherwise, without the prior permission of Springer Publishing Company, Inc.

Springer Publishing Company, Inc.
536 Broadway
New York, NY 10012

89 90 91 92 93 / 5 4 3 2 1

Library of Congress Cataloging-in-Publication Data

The course of later life: research and reflections.
Vern L. Bengtson, K. Warner Schaie, editors.
p. cm.
Bibliography: p.
Includes index
ISBN 0-8261-6220-7
1. Aging—Psychological aspects. 2. Aged—Psychology.
3. Cognition in old age. I. Bengtson, Vern L. II. Schaie, K. Warner (Klaus Warner), 1928-
BF724.8.C68 1989
155.67—dc19 89-4322
CIP

Printed in the United States of America

Contents

James E. Birren

Preface

To think of the later years of life as having a "course"—a set of complex developmental movements, not just a downward trajectory in function or competence—is a relatively new idea in human experience. The research compiled in this volume reflects current multidisciplinary approaches to this new perspective on aging. At the same time, it celebrates the career of one of the foremost contributors to the idea of development in aging, James E. Birren, on the occasion of his retirement as Dean of the Andrus Gerontology Center of the University of Southern California.

Prior to the mid-twentieth century, most characterizations of aging in Western thought reflected the theme of inevitable and irreversible loss. Most people, if they survived infanthood, moved from a few years of prepubescence to young adult and midlife years filled with reproductive and economic activity and then to an often short period of decline and senescence prior to death. Within the life span of those now above the age of 65 however, two developments have occurred to change forever our notions of the course of later life.

The first is a consequence of basic changes in human demographics related to age and social activity. During the past century there has been an astonishing increase in expectable longevity for most members of industrial societies. In the United States alone, life expectancy for females has increased from 41 years for those born in 1900, to almost 80 for those born in 1980. For American males, "retirement" as a period of life was relatively rare in 1900—only 22% of those still living after age 65 were out of the work force; by 1980, 89% of those age 65 and older were no longer working, and men at age 60 could look forward to an average of 15 more years of life. Moreover, the age composition of industrialized societies has altered dramatically in the past century: in 1980 one out of nine Americans was over the age of 65, compared to one in 22 in 1900.

The second development involves the increase in scientific activity and public awareness regarding the later years of life. Prior to 1940 there was relatively little scientific interest in charting changes in function or competence after the middle years and only sporadic public concern about the aged population. There were few studies to support either descriptive or explanatory generalizations about the course of human aging. Within the past few decades, however, there has been an exponential increase in research regarding both pathological and normal changes with age. This growing body of scientific evidence has led to an awareness of the significant diversity among individuals in their responses to biopsychological processes of aging and of the heterogeneity among social aggregates in the aging population. Above all, multidisciplinary research has contributed to the notion of a complex trajectory of human development past the middle years, seen in both decrements and adjustments to biological, psychological, and social competencies after young adulthood.

The purpose of this volume is to present an overview of some current issues and perspectives reflecting scientific analysis of the course of later life. In doing so, the contributing authors present their tribute to James E. Birren, their mentor, colleague, and friend.

In reflecting on a long and productive career such as Birren's there is an opportunity to explore the influences shaping the individual and, in turn, the influence of the individual on the careers of other persons. Birren's graduate studies in psychology began in the summer or 1941 when the clouds of war were already gathering. His studies at Northwestern University were, in fact, heavily influenced by the tempo of the war effort. He found himself working with a team of psychologists and physiologists studying issues of fatigue and drug actions. This led to research in the Mojave Desert during wartime, exploring issues of fatigue in tank drivers who were engaged in extended maneuvers. Proceeding from this experience, he joined the staff of the Naval Medical Research Institute in 1943 as a Junior Naval Officer (Ensign). The pace of research was intense and, almost always, the research team was an interdisciplinary one. In 1947, Birren's professor admitted that Birren had done more research in that six-year period than the professor himself had done in his entire career.

The orientation toward interdisciplinary research was also strengthened from 1947 through 1950 when Birren was at the Gerontology Research Center in Baltimore and subsequently, from 1950 to 1953, when he was assigned by the Public Health Service to the University of Chicago. His papers from that period reflect a decided interest in interscientific issues. For two years he was in residence as a Research Associate at the level of Assistant Professor in the Department of Anatomy at the University of Chicago. That department was interested in the functional characteristics of anatomical structures and had a tradition of having psychologists work with the department. There seems little doubt that this period crystallized Birren's belief that problems of aging have no disciplinary boundaries and, in fact, do not care which discipline solves them.

The period of time at the National Institute of Mental Health (NIMH) as Chief of the Section on Aging, 1953–1964, showed a flowering of interdisciplinary commitment to studies of aging. That collaborative effort led to the Human Aging Study of the NIMH. Birren ended his National Institutes of Health career as head of the Division of Aging of the National Institute of Child Health and Human Development and came to head the Gerontology Program at USC in 1965.

At USC he continued his encouragement of interdisciplinary research on aging and sought to expand it in two ways. One was to strengthen ties among the different professional areas to see that knowledge gathered in research could be applied throughout the professional specialties, and the other was to encourage greater effort within the confines of the disciplines themselves. Thus, significant scholars who helped shape his ideas during his career came from a wide range of backgrounds; they included Isadore Gersh, histologist; Nathan Shock, physiologist; David Solomon, endocrinologist; Malcolm Bick, opthalmologist; Martin Grotjahn, psychoanalyst; Robert Butler, psychiatrist; Louis Sokoloff, physiologist; William Bondareff, neuroanatomist; Jack Botwinick, psychologist; Robert Aldrich, pediatrician; Warner Schaie, psychologist; and others. It would be interesting to trace the ideas that influenced Birren during his long career and to identify them with particular persons of origin. Certainly there is a core of commonality which espoused a lack of dogma, encouraged curiosity, and engendered cooperation.

The answer to the question, "Why did Birren get interested in doing research on aging?" is not readily apparent. His entire career has been devoted to research on aging. The question must have a multifactorial answer of the type of which he is fond. When he began his career, aging was a field needing exploration and explanation. It was a growing area of human concern, and he could learn from cooperating colleagues. Even he is not certain what originally pushed the button that led him to devote over 40 years to doing and encouraging research on aging.

Betty Birren, his wife, collaborated with him and other colleagues in developing gerontology from time to time. During the early years of the Gerontology Research Center in Baltimore, she undertook the organization of the Shock Bibliography on Aging. The field of gerontology was just beginning and she was employed by Nathan Shock to pull together literature on the subject. She went off to what was then the Army Surgeon General's library to search out what was known about aging. She subsequently did another search in the same vein for the Administration on Aging. That was before the automated data systems of present time, and manual searches had to be made among heavy, dusty volumes.

Her organizational skills were also apparent in her activity in the Division of the Behavioral and Social Sciences of the Gerontological Society and in the Division on Adult Development and Aging of the American Psychological Association. One interesting fact is that Betty Birren was the President of the Division of Adult Development and Aging in 1981, 25 years after James Birren had been President. In between those years, three children were raised; among her other responsibilities this seemed to settle upon her easily and she met all her obligations with grace and dispatch.

All of the contributors to this volume have been associated with Birren for substantial periods of time; Jack Botwinick, with whom he began collaborating in 1948, has worked with him the longest of all. Shortly after that he met K. Warner Schaie, who was a graduate student during the 1955 meeting of the American Psychological Association. They spent some time together, junior and senior, exploring some of the psychological issues of aging.

Walter Cunningham, Diana Woodruff-Pak, Robert Butler, Vern Bengtson, Caleb Finch, and Powell Lawton have all contributed at one point or another to the development of Birren's thinking about puzzles of aging of living systems. There is no doubt that he affected most, if not all, of these persons at critical times, and sparked ideas that continue to percolate.

Apparently, the fire of curiosity has not died down in him. One of his current pursuits concerns issues of human attention and aging: he is involved with Dr. Joan McDowd in a project to explore the symbiotic relationship of spouses in covering the attentional and memory lapses of each other. The implicit idea here is that to some extent people are yoked computers and part of their memory store is mutually interdependent. He is interested in the interpersonal strategies for managing the daily lapses of attention when they occur.

His work on autobiography has been going on for about twelve years and he frankly admits he cannot explain how that relates to his work in the psychophysiology of aging. He shrugs his shoulders and simply says, "I'm just interested in it—curiosity is its justification." He is also attempting to develop a plan for California in anticipation of the year 2020 when the population over age 65 will double to over six million persons. He believes we have been acting like ostriches in the face of this mammoth increase that will change the complexion of services, products, and the curriculum of higher education.

James Birren has had a lasting effect on research on aging through his encouragement of collaborative effort, his interdisciplinary orientation, and his filling the Zeitgeist's need for leadership. He has had an impact on foundations, the public, and his colleagues and students.

Birren believes that at the present time gerontology is largely devoid of theory, that it is heavily empirical in its emphasis, and that we are likely to see in the near future an upsurge of theoretical activity. He concludes, "Apparently, there seems to be the conviction that good methods and good design in research automatically imply a good question. That is not so. We need to push for more good questions." Many of his colleagues and former students, a sample of whom contributed to this volume, *are* asking good questions.

We would like to acknowledge the contributions of several individuals in the preparation of this volume. Our greatest debt is to Pauline Robinson, Ph.D., who served as the managing editor and who provided much continuity in moving from draft to final copy of these chapters. She has been an associate of James Birren for many years, and has provided guidance to many programs at the Andrus Gerontology Center, where she served as the UPS Research Professor from 1981 to 1985. We also acknowledge the assistance of Ed Schneider, who followed James Birren as Dean of the School of Gerontology at USC; David Peterson, the Director of the Leonard Davis School of Gerontology and Associate Dean of the Center; and Susan Neuwirth who was instrumental in raising funds for the James E. Birren Chair in Gerontology at USC. We also wish to acknowledge the assistance of Linda Hall who coordinated the manuscript preparation for the entire volume.

VERN L. BENGTSON
K. WARNER SCHAIE

Contributors

James E. Birren, Ph.D., is Professor of Gerontology and Psychology at the Andrus Gerontology Center of the University of Southern California, Los Angeles. Currently he is a Brookdale Distinguished Scholar and has research interests in aging and attention and in the process and analysis of autobiographies.

Jack Botwinick, Ph.D., is Professor Emeritus in the Departments of Psychology and Neurology, Washington University, St. Louis, Mo. He has a long history in research on aging, with emphasis in psychomotor and cognitive functions. The main focus of his work during the past decade has been in research on Alzheimer's disease.

Robert N. Butler, M.D., has been Brookdale Professor and Chairman of the Gerald and May Ellen Ritter Department of Geriatrics and Adult Development at Mount Sinai Medical Center, N.Y., since 1982. He was the founding director of the National Institute on Aging in 1976. Winner of a Pulitzer Prize in 1976 for his book *Why Survive: Being Old in America,* Dr. Butler is editor-in-chief of *Geriatrics,* a journal for primary care physicians. A member of the Institute of Medicine of the National Academy of Sciences, he is a founding Fellow of the American Geriatrics Society.

Walter R. Cunningham, Ph.D., is Professor of Psychology at the University of Florida, Gainesville. He has authored numerous papers on the topic of changes in intellectual functioning with age and co-authored a textbook on gerontology. He is currently studying longitudinal patterns of change in various cognitive and intellectual abilities in the elderly.

Caleb E. Finch, Ph.D., is the ARCO/William F. Kieschnick Professor in the Neurobiology of Aging at the Andrus Gerontology Center of the University of Southern California, Los Angeles. His major interests are the mechanisms of brain aging and the identification of pacemaker systems in mammalian aging.

M. Powell Lawton, Ph.D., has been Director of Behavioral Research at the Philadelphia Geriatric Center for the past 24 years, and is also an

Adjunct Professor of Human Development at the Pennsylvania State University and a Research Scientist at Norristown State Hospital. He has done research in the environmental psychology of later life, in assessment of the aged, the psychological well-being of older people, and evaluative studies of programs for the aged and for the mentally ill. Dr. Lawton is a past President of the American Association's Division on Adult Development and Aging. He is the author of *Environment and Aging* and *Planning and Managing Housing for the Elderly.* A past president of the Gerontological Society of America, he is the first editor of the American Psychological Association's journal, *Psychology and Aging.*

Daniel F. Schmidt, Ph.D., is currently Assistant Professor of Psychology at the University of Maine at Presque Isle.

Diana S. Woodruff-Pak, Ph.D., is Professor of Psychology at Temple University, Philadelphia, Pa. Her most recent book is *Psychology and Aging,* and she is President-elect of Division 20 (Adult Development and Aging) of the American Psychological Association. Her primary research interest is the neurobiology of learning, memory, and aging.

The Course of Later Life
Research and Reflections

1

The Brain, Genes, and Aging

Caleb E. Finch

James Birren, in my view, was a fundamental catalyst for the birth and continued growth of the neurobiology of aging. Three decades ago, together with Patrick Wall, he did a pioneering study to characterize the effects of aging on a peripheral nerve tract from rats, using the most advanced anatomic and electrophysiological techniques available at the time. Birren was then temporarily located at the University of Chicago pending the preparation of suitable facilities at the National Institutes of Health in Bethesda. The work showed clearly that the sciatic nerve axons maintained their numbers of cells and electrophysiological (cable conduction) properties throughout the rat's life span (Birren & Wall, 1956). This study unambiguously pointed to synapses as the most likely sites for the age changes which cause slowing of brain functions.

Subsequent studies from many labs, including mine, have described age-related changes in synaptic mechanisms, including decreased numbers of some receptors and neurotransmitter turnover, particularly for the catecholamines (Finch, 1973; Rogers & Bloom, 1985). Moreover, the Birren and Wall (1956) study made it clear that not all types of nerve cells are lost during aging. This was the first study that I know of to evaluate both the electrical and anatomical properties of nerve cells as a function of aging.

These conclusions might have seemed discouraging to most scientists at the time because this study failed to show an effect of age on fundamental properties of neurons. Today we have abundant supporting data which demonstrate that in the absence of specific neurodegenerative diseases such as Alzheimer's the numbers of neurons in most peripheral and central pathways, including the cerebral cortex, are not reduced during aging (Haug et al., 1983; Terry, DeTeresa, & Hansen, 1987). A trend for large neurons to shrink might have contributed to the conclusion of massive neuronal loss during human aging, a conclusion that many of us no longer accept. Thus, Burns's (1958) famous estimate that adult humans lose 100,000 neurons each day can no longer be supported.

However controversial the present literature is, it can trace one of its firmest foundations back to the Birren and Wall paper (1956). I vividly recall reading this paper while a graduate student at Rockefeller University in 1965. Birren's outstanding book, *The Psychology of Aging* (1964), which cited the paper, serendipitously came to my attention while I was browsing in the library of the Rockefeller University. I was thus led to consider the tremendously exciting possibility that not all nerve cells might be genetically programmed to undergo senescent deterioration simply because of their postmitotic (nondividing) status. Some of you will recall the impact that Hayflick's (1965) studies on the "in vitro senescence" of human diploid fibroblasts were having on biological studies of aging. At that time, and still to some extent today, the postmitotic status of fibroblasts and other cells was considered to be a characteristic of senescence. This is a perspective that I seriously began to doubt from the time I read the Birren and Wall paper.

It can be said now quite confidently that merely because a nerve cell is postmitotic and nondividing, it does not inevitably undergo age-related deterioration, at least within the present human life span. A good example is given by the undiminished capacity of healthy older men to secrete the neuropeptide vasopressin in response to increased blood osmolality (Helderman et al., 1978); vasopressin is made by hypothalamic neurons and is secreted directly into the blood. Although no direct measurement of the number of these neurons has been made across the adult life span, the hormone data give no hint of deteriorating functionality. Moreover, if the diffuse and haphazard neuronal damage caused by stroke and other common age-related vascular impairments is treated separately, most evidence points to a high selectivity in the deterioration of particular brain cells during the course of aging. The task at hand then is to understand better what factors lead to the deterioration of brain cells in some parts of the brain, and it is here that genetic influences become a major consideration.

ALZHEIMER'S DISEASE AND HUNTINGTON'S CHOREA

At least two neurological diseases whose incidence increases during age have genetic components: Alzheimer's disease, which rarely occurs before midlife and which has a rapidly increasing incidence after 70 years, and Huntington's disease, which most commonly occurs during midlife (see Figure 1.1). First, Alzheimer's disease is now recognized to include a significant influence from family background, which some investigators have recently estimated may account for the majority of cases (Heston, Mastin, Anderson, & White, 1981; Heyman et al., 1983; Mohs, Breitner, Silverman, & Davis, 1987; Tanzi et al., 1981). It is, of course, difficult to establish hereditary influences on a disease that has such a late age of onset because there are many causes of death unrelated to Alzheimer's disease that could cause an individual to die before he or she manifested the disease. The selectivity of neuronal damage during Alzheimer's disease has recently become particularly clear. Only in select pathways and regions do neurons degenerate (Kemper, 1984; Rogers & Morrison, 1985). The cerebellum, for example, which contains nearly half of the cells in the brain, is untouched during Alzheimer's disease. It is mainly in select areas of the frontal cortex and subcortical areas that cell deterioration occurs during Alzheimer's disease. Characteristic manifestations include intraneuronal neurofibrillary tangles and the formation of neuritic or senile plaques involving groups of adjacent nerve fibers.

Thus, the genes that may in the future be identified to cause Alzheimer's disease do so in an exquisitely selective way, touching only some cells and leaving others in very nearby regions completely normal in appearance and function. One may thus view Alzheimer's disease as an example of selective gene regulation, in which the apparent genetic trait is selectively expressed to impair just some neurons. It is an index of the remarkable subtlety of neurons that many other cells remain normal despite their nearly identical properties in the younger individual in regard to morphology and neurochemistry.

A different example of selective neuronal deterioration under genetic control is presented by Huntington's chorea, a rare genetic disease that has autosomal dominance and that is typically first manifested during midlife (see Figure 1.1). Just one copy of the harmful gene from either

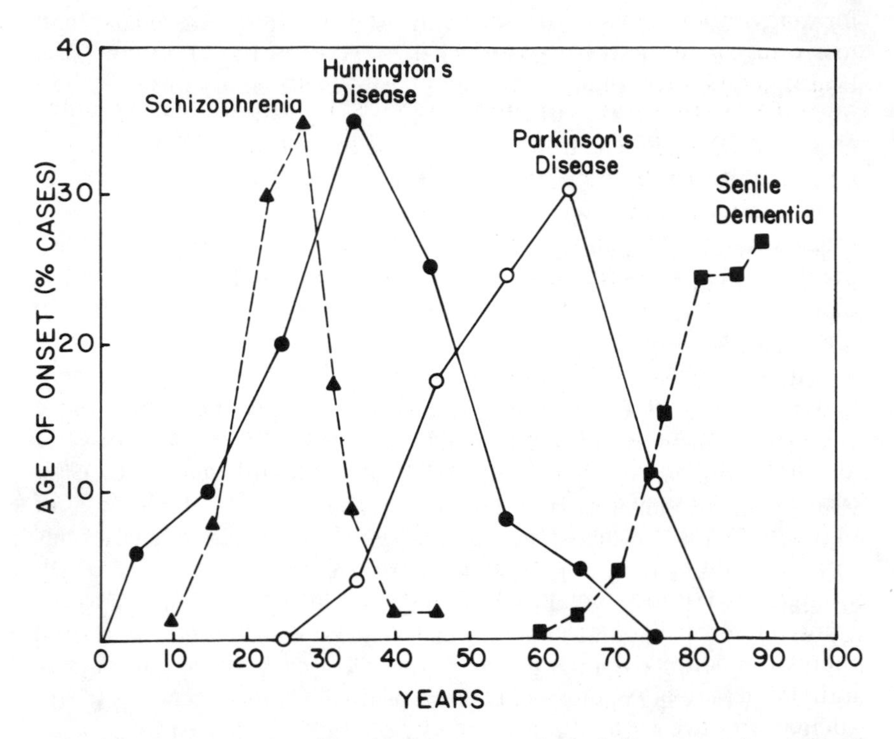

FIGURE 1.1 The age distribution of four neurological diseases at the onset of symptoms. Original souces cited in Morgan, May, and Finch (1987). More recent data indicate that the incidence of Parkinsonism continues to rise at later years, contrary to the figure.

parent condemns its carrier to a fatal dementing illness that damages a different set of cells in the basal ganglia than are altered during Alzheimer's disease. Both Alzheimer's disease and Huntington's chorea share a major similarity in that the adverse gene is not expressed until long after development. Individuals who carry the Huntington's gene are apparently normal until their twenties or thirties. Thereafter, they show a highly selective deterioration of nerve cells that kills them in about 10 years.

The gene for Huntington's chorea has recently been located in a small region of human chromosome 4 (Gusella et al., 1983). Within the foreseeable future, the offending DNA sequence itself will probably be cloned and its base sequence and organization deciphered. It will then be possible to study the factors that influence its expression. For example, the gene might be injected into mouse embryos where, as shown for many other genes, it may be incorporated into the germ line DNA to produce a new genetic species of the mouse carrying a human gene. I have called such genetic transformants "hu-mice" because they are mice that carry a chunk of human DNA spliced into their own chromosomal DNA (Finch, 1985). Once the Huntington's gene is placed in a mouse, it may be then possible to learn the factors that control its expression. A spectacular example of this approach is given by the introduction of the human insulin gene into mice; in a few transformants, the expression of the foreign insulin gene was regulated by the blood levels of glucose and other physiological controls (Selden, Skoskiewicz, Howie, Russell, & Goodman, 1986). Thus, we should be able to analyze in short-lived mice the factors that control many human genes that may in the future be implicated in adverse aspects of aging.

We already suspect the existence of factors controlled by one set of genes that influence the activities of other genes during aging. For example, the age of onset of Huntingtonism is correlated with the life span of unaffected family members. This finding stemmed from observations I made about the many similarities between the cellular and chemical changes in Huntingtonism and those which normally occur during aging (Finch, 1980). I speculated that the onset of Huntingtonism must then be in some way closely linked to brain aging processes and that individual variations in onset age might reflect genetic polymorphisms, such as we have found in the neurochemistry of inbred rodents (Severson, Pittman, Gal, Molinoff, & Finch, 1986; Severson, Randall, & Finch, 1981). In turn, Farrer and his colleagues examined families with different ages of onset of Huntingtonism and discovered an important relationship of the onset age to the longevity on unafflicted family members (Farrer, Conneally, & Yu, 1984). Thus, early-onset cases come from short-lived families, while late-onset cases come from long-lived families (see Figure 1.2).

It is possible that we can identify the factors made by other genes that influence the expression of the gene for Huntingtonism. The contemporary models of genes in higher cells include regulatory sequences that are sensitive to different "trans-acting" factors in the cell environmental factors, including hormones and growth factors, which modulate or influence the transcription of the gene itself (Lewin, 1985). Thus, we can consider that the cellular and organismic environment of a gene may influence its expression.

In the future it may be possible to manipulate the environment of a gene for Huntington's chorea or even Alzheimer's disease to reduce its adverse effects. One example of this is suggested by a different hereditary disease, phenylketonuria (PKU), which also causes irreversible

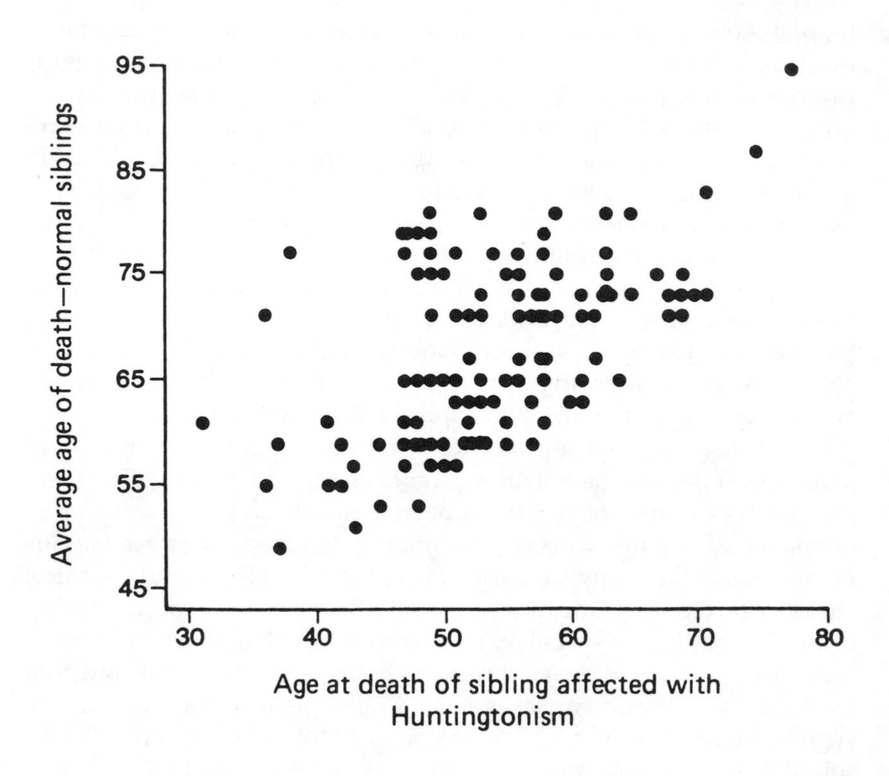

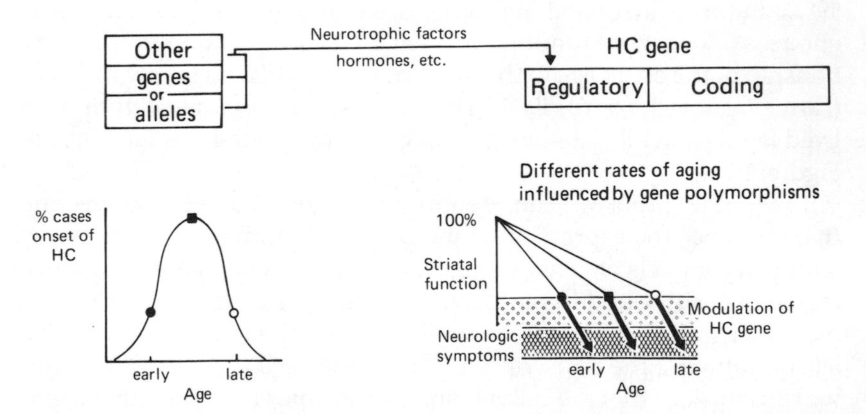

FIGURE 1.2 Hypothesis relating the age at death in Huntington's disease (HD) to the longevity of unafflicted family members (noncarriers). Figure 1.2a panel is data of Farrer et al. (1984). Figure 1.2b suggests that other genes may influence transcription of the HD locus via trans-acting factors. See Finch (1987) for details.

brain damage in children. This disease used to condemn tens of thousands of children in the United States to a life in institutions for the feebleminded. However, it was discovered that carriers of this gene had an enzymatic defect so that the usual diets elevated blood levels of phenylalanine and caused irreversible brain damage. Merely by reducing the levels of these amino acids in the diet it was possible to protect carriers of this genetic defect during their critical childhood years, and so many of them have now grown up across the United States to lead far more productive lives without as major mental impairment. This heartening example suggests the potential for controlling other genetic defects that cause otherwise irreversible brain damage, possibly even Alzheimer's disease.

It may be possible to learn how to control adverse effects of genes by mild environmental interventions long before the gene has adverse effects. So far, of course, there are no approaches like this that have been applied to Alzheimer's disease or Huntingtonism, but as we identify carriers of hereditary defects that lead to late-onset neurological disorders it may be possible to recognize environmental influences that modify the course of the disease. In the future these discoveries might lead to prevention of the adverse effects.

HORMONALLY INDUCED DAMAGE

Hormonally induced neuronal damage may underlie neurological deterioration in some individuals. Diabetes has long been known to increase the risk of blindness (diabetic retinopathy), slowed peripheral nerve conduction velocity, and impotence. A range of neuronal degeneration can also be induced by experimental diabetes in laboratory rats, including slowing of axonal transport (Medori, Autilio-Gambetti, Monaco, & Gambetti, 1985), deterioration of peripheral motor tracts (Sima & Robertson, 1979), and damage to hypothalamic centers of endocrine control (Bestetti, Locatelli, Tirone, Tossi, & Muller, 1985). Chronic elevations of steroids in rodents can also cause long-lasting neuronal damage that includes effects of estrogens on the hypothalamus and pituitary (Finch, Felicio, Mobbs, & Nelson, 1984; Kohama, May, & Finch, 1986; Schipper, Brawer, Nelson, Felicio, & Finch, 1981) and of corticosteriods on the hippocampus (Finch & Landfield, 1985; Sapolsky, Krey, & McEwen, 1985). The changes are selective for particular neurons and functions and can be included in a new category of aging phenomena that I think of as "nonischemic neuronal death." The possibility that stress steroids can kill neurons in adult rodents (Sapolsky et al., 1985) supports the long-standing view that stress may contribute to aging (Klonoff, McDougall, Clark, Kramer, & Horgon, 1976; Selye & Prioreschi, 1960; Thygesen, Hermann, & Willanger, 1970).

Until future studies have demonstrated irreversible effect from steroids on primate brain neurons, we cannot presume the validity of any inferences drawn from rodents to humans (Finch, 1986). Nonetheless, the selectivity of these effects on rodents and their similarity to many spontaneous changes of aging is consistent with a role of factors extrinsic to neurons and other cells in age-related damage (Finch, 1976; Finch, Felicio, Mobbs, & Nelson, 1984; Finch, Foster, & Mirsky, 1969).

EFFECTS OF STRESS

We are currently analyzing the effects of stress on messenger RNA levels in the hippocampus in rats (Nichols, Lerner, Masters, May, Millar, & Finch, 1988) in order to define the role of gene regulation in the transition between reversible and irreversible effects of stress. These studies are part of our ongoing investigation of the endocrine-sensitive centers of the brain and their relation to possible neuroendocrine cascades of aging, which I have proposed play a major role in regulating cellular changes throughout the body during aging through changes in hormones and neurosecretions (Finch, 1973, 1976; Finch et al., 1984). Here again I can return to one of Birren's prescient insights:

> Thus, if aging is manifested by the endocrines, it is probably a reflection of changes in their stimulation, particularly the pituitary, and ultimately from the hypothalamus and higher brain centers. The distinction here is one of primary and secondary aging. (Birren, 1964, p. 53)

VARIABILITY IN NUMBERS OF NEURONS

Besides the possibility of individual patterns of brain aging that may result from individual differences in the exposure to stresses, individual differences in aging may also arise from differences in the numbers of neurons and other irreplaceable cells that are acquired only during embryonic development. My first introduction to this realm of possibilities was through Birren's superb synthesis of facts and theory in the volume he assembled, *Handbook of Aging and the Individual* (1959). This book contains many examples of the remarkable variability in the outcomes of aging between individuals. This variability could arise from many different sources both hereditary and environmental at any point after conception. One of the sources of this variability, I believe, can be traced to the pattern of development which may lead individuals to have varying numbers of cells in different parts of their bodies and their brains. Even in highly inbred mice which are genetically identical, a remarkable range of cell numbers can be seen. For example, the number of egg cells in the ovary can be seen to vary between individual mice by as much as two-fold (Gosden, Laing, Felicio, Nelson, & Finch, 1983).

The relatively limited amount of data that give detailed counting of cells in adult humans of reliable health status before death suggests that there are substantial variations among individuals in the numbers of egg cells of the ovary as well as in the numbers of neurons, variations which are in the range of 30% or more in the brain regions and nerve tract. The cited papers do not all have optimal numbers of young healthy adults but nonetheless suggest the considerable extent of neuronal variations between adults (Hoffman & Schnitzlein, 1961; Swaab & Fliers, 1985; Tomonaga, 1983; Vijayashankar & Brody, 1979). How then might such a range of variations arise? Recent studies in fish and amphibia indicate that the origin of neurons during development is subject to considerable randomness. In the predecessors of neurons at each cell division there is an approximate probability that the division products will become neurons, will differentiate into other cell types, or will die (see Figure 1.3). This considerable randomness in cell fate, when summated over the extensive number of divisions that take place during development, leads to individual variation in the number of cells in the adult nervous system (Jacobson & Moody, 1984; Winklbauer & Hausen, 1983). This phenomenon has been demonstrated now for three different types of neurons.

If such a phenomenon also applies to mammals (there is no reason to suspect it does not), then individuals will be born with varying numbers of neurons in different parts of their brain. As a consequence, one can consider that individuals differ in the numbers of reserve cells needed to serve critical functions in their brains. Thus, some individuals might be given at birth an excess of neurons in some parts of the brain and a minimum number in other parts needed for normal function. Since neurons cannot be replaced in the adult animal, I propose then that individual variations in the numbers of neurons might account for the diversity among individuals in the outcome of their aging; thus, some individuals might be buffered for loss of cells that occurs as a function of stroke or other accident in some parts of their brain but not in others. A major question in the neurobiology of aging, then, con-

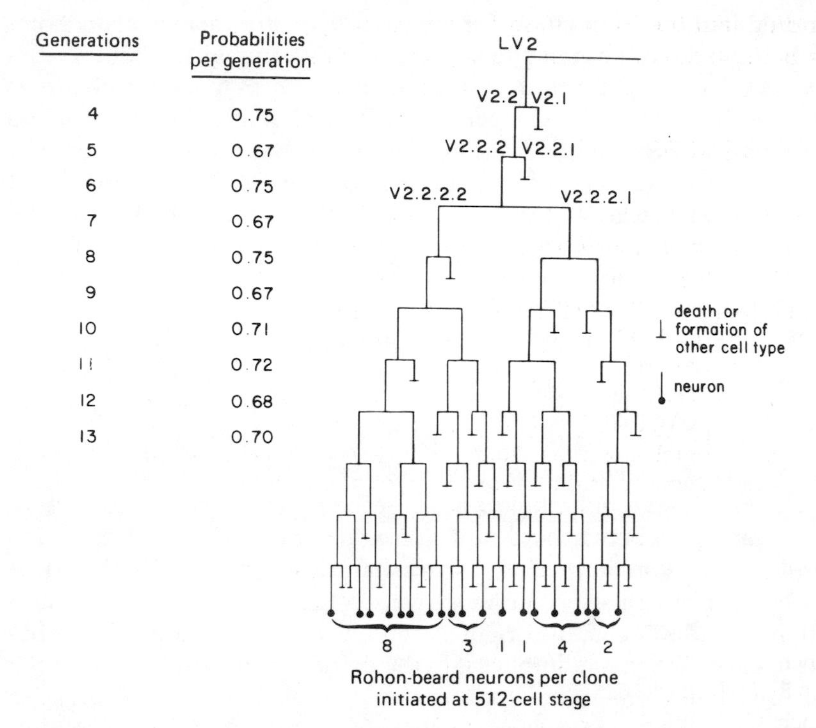

FIGURE 1.3 The clonal lineage of neurons (Rohon-Beard neurons) during development of the frog *Xenopus*, showing that the probability of continuing in the clonal lineage is constant across the initial cell generations. This probability means that the numbers of neurons will vary extensively for each clone at the termination of each lineage when development is complete (Jacobson & Moody, 1984; Winklbauer & Hausen, 1983). The variations in neuron number fit nicely into the Poisson distribution. If such processes occur in mammals, there could be considerable variation between individuals in the numbers of neurons in different brain regions, and this could alter the impact of neuron loss during aging or neurological diseases. (Redrawn from Jacobson, 1985.)

cerns the role of developmentally random variations in cell number on the later course of life.

Thus, we can consider that aging involves an interplay of genetics and random factors such that for each individual there may be a unique fingerprint of that person's destiny. The origins of this subject, in my view, can be traced back to Birren's early work in the neurobiology of aging, which showed that neuronal number and function for at least some cell types did not inevitably decline with age. This fundamental finding laid the foundation for my research, and I always regarded it as being one of the most basic contributions in our subject.

REFERENCES

Bestetti, G., Locatelli, V., Tirone, F., Tossi, G. L., & Muller, E. E. (1985). One month of streptozotocin-diabetes induces different neuroendocrine and morphological alterations in the hypothalamo-pituitary axis of male and female rats. *Endocrinology, 117,* 208–216.

Birren, J. E. (Ed.). (1959). *Handbook of aging and the individual.* Chicago: University of Chicago Press.

Birren, J. E. (1964). *The psychology of aging.* Englewood Cliffs, NJ: Prentice-Hall.

Birren, J. E., & Wall, P. D. (1956). Age changes in conduction, velocity, refractory period, number of fibers, connective space and blood vessels in sciatic nerve of rats. *Journal of Comparative Neurology, 104,* 1–16.

Burns, B. D. (1958). The mammalian cerebral cortex. In E. Arnold (Ed.), *Monograph of the Physiological Society,* No. 5.

Farrer, L. A., Conneally, P. M., & Yu, P-I. (1984). The natural history of Huntingtons disease: Possible role of "ageing genes." *American Journal of Medical Genetics, 18,* 115–123.

Finch, C. E. (1973). Catecholamine metabolism in the brains of aging male mice. *Brain Research, 52,* 261–276.

Finch, C. E. (1976). The regulation of physiological changes during mammalian aging. *Quarterly Review of Biology, 51,* 49–83.

Finch, C. E. (1980). The relationships of aging changes in the basal ganglia to manifestations of Huntington's chorea. *Annals of Neurology, 7,* 406–411.

Finch, C. E. (1985). Modulation of aging processes in the brain. In M. Bergener, M. Ermini, & H. B. Stahelin (Eds.), *Thresholds in aging* (pp. 175–188). New York: Academic Press.

Finch, C. E. (1986). New questions about steroids. *Journal of the American Geriatric Society, 34,* 393–394.

Finch, C. E. (1987). Neural and endocrine determinants of senescence: Investigation of causality and reversibility by laboratory and clinical interventions. In H. R. Warner (Ed.), *Modern biological theories of aging* (pp. 261–306). New York: Raven Press.

Finch, C. E., Felicio, L. S., Mobbs, C. V., & Nelson, J. F. (1984). Ovarian and steroidal influences on neuroendocrine aging in female rodents. *Endocrine Reviews, 5,* 467–497.

Finch, C. E., Foster, J. R., & Mirsky, A. E. (1969). Ageing and the regulation of cell activities during exposure to cold. *Journal of General Physiology, 54,* 690–712.

Finch, C. E., & Landfield, P. W. (1985). Neuroendocrine and autonomic functions in aging mammals. In C. E. Finch & E. L. Schneider (Eds.), *Handbook of the biology of aging* (pp. 567–594). New York: Van Nostrand Reinhold.

Gosden, R. G., Laing, S. C., Felicio, L. S., Nelson, J. F., & Finch, C. E. (1983). Imminent oocyte exhaustion and reduced follicular recruitment mark the transition to acyclicity in aging C57BL/6J mice. *Biology of Reproduction, 28,* 255–260.

Gusella, J. F., Wexler, N. S., Conneally, P. M., Naylor, S. L., Anderson, M. A., Tanzi, R. E., Watkins, P. C., Ottina, K., Wallace, M. R., Sakaguchi, A. Y., Young, A. B., Shoulson, I., Bonilla, E., & Martin, J. B. (1983). A polymorphic DNA marker genetically linked to Huntington's disease. *Nature, 306,* 234–238.

Haug, H., Barnwater, U., Eggers, R., Fischer, D., Kuhl, S., & Sass, N. L. (1983). Anatomical changes in aging brain: Morphometric analysis of the human prosencephalon. In J. Cervos-Navarro & H-I Sarkander (Eds.), *Brain aging: Vol. 21. Neuropathology and neuropharmacology* (pp. 1–12). New York: Raven Press.

Hayflick, L. (1965). The linked in vitro lifespan of human diploid cell strains. *Experimental Cell Research, 37,* 614–636.

Helderman, J. H., Vestal, R. E., Rowe, J. W., Tobin, J. D., Andres, R., & Robertson, G. L. (1978). The response of arginine vasopressin to intravenous ethanol and hypertonic saline in man: The impact of aging. *Journal of Gerontology, 33,* 38–47.

Heston, L. L., Mastin, A. R., Anderson, E., & White, J. (1981). Dementia of the Alzheimer type: Clinical genetics, natural history, and associated conditions. *Archives of General Psychiatry, 38,* 1085–1090.

Heyman, A., Wilkinson, W. E., Hurwitz, B. J., Schmechel, D., Sigmon, A. H., Weinberg, T., Helms, M. J., & Swift, M. (1983). Alzheimer's disease: Genetic aspects and associated clinical disorders. *Annals of Neurology, 14,* 507–515.

Hoffman, H. H., & Schnitzlein, H. N. (1961). The numbers of nerve fibers in the vagus nerve of man. *Anatomical Record, 139,* 429–434.

Jacobson, M. (1985, April). Clonal analysis of the vertebrate CNS. *Trends in Neuroscience,* 151–155.

Jacobson, M., & Moody, S. A. (1984). Quantitative lineage analysis of the frog's nervous system: I. Lineages of Rohon-Beard neurons and primary motoneurons. *Journal of Neuroscience, 4,* 1361–1369.

Kemper, T. (1984). Neuroanatomical and neuropathological changes in normal aging and in dementia. In M. L. Albert (Ed.), *Clinical neurology of aging* (pp. 9–52). New York: Oxford University Press.

Klonoff, H., McDougall, G., Clark, C., Kramer, P., & Horgon, J. (1976). The neuropsychological, psychiatric, and physical effects of prolonged and severe stress: 30 years later. *Journal of Nervous and Mental Disorders, 163,* 246–252.

Kohama, S. G., May, P. C., & Finch, C. E. (1986). Oral administration of estradiol-induced age-like reproductive acyclicity in C57BL/6J mice. *Neuroscience Society Abstracts, 12,* 1466.

Lewin, B. (1985). *Genes II.* New York: Wiley.

Medori, R., Autilio-Gambetti, L., Monaco, S., & Gambetti, P. (1985). Experimental diabetic neuropathy: Impairment of slow transport with changes in axon cross-sectional area. *Proceedings of National Academy of Science USA, 82,* 7716–7720.

Mohs, R. C., Breitner, J. C. S., Silverman, J. M., & Davis, K. L. (1987). Alzheimer's disease: Morbid risk among first degree relatives approximates 50% by 90 years of age. *Archives of General Psychiatry, 44,* 405–408.

Morgan, D. G., May, P. C., & Finch, C. E. (1987). Dopamine and serotonin systems in human and rodent brain: Effects of age and neurodegenerative disease. *Journal of the American Geriatric Society, 35,* 334–345.

Nichols, N. R., Lerner, S. P., Masters, J. N., May, P. C., Millar, S. L., & Finch, C. E. (1988). Rapid and select increases in poly(A)—containing RNA sequences in rat hyppocampus during corticosterone treatment. *Molecular Endocrinology, 2,* 284–290.

Rogers, J., & Bloom, F. E. (1985). Neurotransmitter metabolism and function in the aging central nervous system. In C. E. Finch & E. L. Schneider (Eds.), *Handbook of the biology of aging* (pp. 645–691). New York: Van Nostrand Reinhold.

Rogers, J., & Morrison, J. H. (1985). Quantitative morphology and laminar distributions of senile plaques in Alzheimer's disease. *Journal of Neuroscience, 5,* 2801–2808.

Sapolsky, R. M., Krey, L. C., & McEwen, B. S. (1985). Prolonged glucocorticoid exposure reduced hippocampal neuron number: Implications for aging. *Journal of Neuroscience, 5,* 1221–1226.

Schipper, H. Brawer, J. R., Nelson, J. F., Felicio, L. S., & Finch, C. E. (1981). Role of the gonads in the histologic aging of the hypothalamic arcuate nucleus. *Biology and Reproduction, 24,* 784–794.

Selden, R. F., Skoskiewicz, M. J., Howie, K. B., Russell, P. S., & Goodman, H. M. (1986). Regulation of human insulin gene expression in transgenic mice. *Nature, 321,* 525–528.

Selye, H., & Prioreschi, P. (1960). In N. W. Shock (Ed.), *Aging, some social and biological aspects* (pp. 261–272). Washington, DC: AAAS (Publication No. 65).

Severson, J. A., Pittman, R. N., Gal, J., Molinoff, P. B., & Finch, C. E. (1986). Genetic influence on the regulation of beta-adrenergic receptors in mice. *Journal of Pharmacology and Experimental Therapeutics, 236,* 24–29.

Severson, J. A., Randall, P. K., & Finch, C. E. (1981). Genotypic influences on striatal dopaminergic regulation in mice. *Brain Research, 210,* 201–215.

Sima, A. A. F., & Robertson, D. M. (1979). Peripheral neuropathy in the diabetic mutant mouse: An ultrastructural study. *Laboratory Investigation, 40,* 627–632.

Swaab, D. F., & Fliers, E. (1985). A sexually dimorphic nucleus in the human brain. *Science, 228,* 1112–1115.

Tanzi, R. E., St. George-Hyslop, P. H., Haines, J. L., Polinsky, R. J., Nee, L., Foncins, J. F., Nevell, R. L., McClatchey, A. I., Conneally, P. M., Gusella, J. F. (1987). The genetic defect in familial Alzheimer's disease is not tightly linked to the amyloid β-protein gene. *Nature, 329,* 156–159.

Terry, R. D., DeTeresa, R., Hansen, L. A. (1987). Neocortical cell counts in normal human aging. *Annals of Neurology, 21,* 530–539.

Thygesen, P., Hermann, K., & Willanger, R. (1970). Concentration camp survivors in Denmark: Persecution, disease, disability, compensation. *Danish Medical Bulletin, 17,* 65–108.

Tomonaga, M. (1983). Neuropathology of the locus ceruleus: A semiquantitative study. *Journal of Neurology, 230,* 231–240.

Vijayashankar, N., & Brody, H. (1979). A quantitative study of the pigmented neurons in the nuclei locus coeruleus and subcoeruleus in man as related to aging. *Journal of Neuropathology and Experimental Neurology, 38,* 490–497.

Winklbauer, R., & Hausen, P. (1983). Development of the lateral line system in Xenopus laevis: II. Cell multiplication and organ formation in the supraorbital system. *Journal of Embryology and Experimental Morphology, 76,* 283–296.

2

Environmental Proactivity in Older People

M. Powell Lawton

Very few people attempting to say something new about the psychology of later life could go far without acknowledging the contribution of James Birren to the idea. This fact is widely recognized for the psychologist writing about sensory-motor processes, cognition, and information processing. It may be less obvious when the topic is personality or the subjective aspects of aging. It is worth noting, however, that wisdom (Clayton & Birren, 1980), affect (Culbertson & Birren, 1973), love (Neiswander & Birren, 1973; Reedy & Birren, 1978), the urban environment (Birren, 1969), competence (Birren, 1985), and the construction of the self (Birren & Hedlund, in press) have also been areas to which Birren's original thinking has been directed. In fact, the term "self-construction" is probably the best overall term to describe his view of how the person functions in life span perspective. Old age is clearly seen by Birren as a time of continuing effort to manage who one is, what one does, and who one will be. Many of us have moved into this area, and it is to such a view of the older person that this chapter will be devoted.

It has become fashionable to demonstrate the "plasticity" of older people, that is, that they can still learn and adapt to new conditions. I should like to redefine plasticity in terms of alternative concepts of what is the object and what is the subject of plasticity. Plasticity is applied to the older person as an object—that is, the older person's behavior remains capable of being shaped by stimulation of the appropriate variety rather than being rigidified and impervious to change through reinforcement or other forms of patterned stimulation. In speaking about person in relation to environment I shall certainly use this concept that the older person remains subject to influence by his or her environmental context.

However, I want to argue equally strongly that environment is just as likely to be the object and the older person the subject of plasticity. The purpose of my comments will be to demonstrate that older people, like others, engage in *transactions* with their environments, a situation that makes it often difficult to distinguish subject from object.

PROACTIVITY–DOCILITY MODEL

This argument is based on a proactivity–docility model, a more general conception of the older person than the one originally advanced in Lawton and Nahemow's (1973) ecological model of aging. This model originally conceived of person and environment as interacting in such a way as to attain a favorable outcome (as indexed by either adaptive behavior or psychological well-being) through maintaining a state of balance between personal competence and environmental press (or demand) level. Adaptation level (Helson, 1964) is a theoretical balance point about which temporally changing discrepancies between competence and press occur. Deviations from adaptation level may occur in the form of mild excesses of environmental press beyond the competence of the moment ("zone of maximum performance potential") and mild deficiencies of press level in relation to competence ("zone of maximum comfort"). Deviations of these magnitudes allow outcomes that are generally favorable. Greater deviations occasion negative behavioral and/or affective outcomes, that is, disorganized or dysfunctional behavior, depression, and other psychopathological symptoms.

One facet of the model was the "environmental docility hypothesis." Briefly, this hypothesis suggested that environment exerts a stronger effect on the person as competence decreases (Lawton, 1970). This conception implied a one-way action of environment on person. "Transaction" is clearly a better term to describe reciprocal interactive relationships between person and environment. This conception is much more in keeping with what we know about aging because it allows the person to act as both initiator and responder in relation to the environment.

Extensions of the model to account for proactive behavior and shaping one's environment were developed (Lawton, 1980; Lawton, in press). In these elaborations the relationships between both externally imposed and self-initiated change in stimulation away from adaptation level, and the resulting affective and motivational states, were developed; this conception was applied specifically to meaningful activities and the uses of time by older people (Lawton, 1985a). That is, the person was seen as being an active agent in regulating the balance among (a) neutral, automatic routine (adaptation level), (b) positive stimulus-level deviations, and (c) negative stimulus deviations. An illustration of the effect of reducing excess press may be found in research on planned housing. A reduction of negative affect in response to self-initiated moves from an overdemanding community environment to a lower-press planned housing environment (Lawton & Cohen, 1974) or to the more protective congregate housing (Lawton, 1976) was reported. Thus, outcomes of stimulus reduction were favorable when the initial press level substantially exceeded competence.

Recent elaboration of this model has led to a preliminary attempt to account conceptually for the changing, dynamic mixes of self-initiated versus other-initiated transactions between person and environments. Rather than seeing only generalized affective (positive or negative) and behavioral (adaptive or maladaptive) outcomes, we see a more complex process where motivation, affect, cognition, and behavior are regulated by varying degrees of action and reaction by the person. While the ultimate outcomes may still be characterized in terms of generalized psychological well-being and adaptive behavior, the mediating circumstances include more specific affects and behaviors that are willfully sought by the person as well as instigated by environmental demands. The duality docility–proactivity is, of course, a continuum, capable of differentiation in both space and time. It is most important to acknowledge that each person maintains a changing mix of proactive and reactive behaviors.

In brief, the relative ascendance of docility (a stance whereby the person responds to external conditions not of his or her own choice or construction) versus proactivity (choice of external conditions according to personal wish or need) is a function of both lifelong preferences and personality ("assertive" and "dependent," "introverted" and "extroverted," and all mixes thereof) and personal competence.

The original docility model used the construct personal competence for heuristic purposes. To conceive of the person in these terms requires one to reduce what is in life a dynamic and transactional process to a static, traitlike entity. As argued at greater length elsewhere (Lawton, 1982), for the present purpose the "basic" competences are the aggregate or the pattern of objectively measured basic biological, sensory, motoric, and cognitive functions, measured as free as possible of influences of environmental context and short temporal variations. Physiological or anatomical health, vision, hearing, and context-free intelligence are the best approximations we have as measures of basic competences.

The docility–proactivity conception suggests that the most successful end states (psychological well-being, adaptive behavior) arise from the person's ability to experience the broadest possible variety of internal and external affects, cognitions, and behaviors, this generalization being conditioned first by lifelong personality preferences and contemporary competence level and second by the extent of self-direction of behavior. The nature of the experience is regulated by both internally and externally imposed factors.

ENVIRONMENTAL PROACTIVITY

The literature relative to the Lawton and Nahemow conception was reviewed (Lawton, 1982) some time ago, and in general it supported the environmental docility hypothesis. More recent research (Carp, 1984; Lieberman & Tobin, 1983) has found it necessary to use the concepts of environmental resources and personal preferences to account for the active and constructive searches older people make for satisfying environmental contexts. For example, Lieberman and Tobin (1983) found that the quality of relocation environments to which deinstitutionalized older mental hospital patients were sent was strongly related to mental health outcome among the most competent patients. Environmental quality was irrelevant to outcome for the least competent. These findings directly contradicted the environmental docility hypothesis.

These data and more general theoretical considerations (Carp, 1984) clearly required an "environmental proactivity hypothesis." This hypothesis suggests that the greater the competence of the person, the more likely the person's needs and preferences will be successfully exercised to search the environment for resources to satisfy the needs. Thus, structuring environment in terms of resources rather than press, a greater proportion of all objectively existing resources in the external environment will, first, be evident to the more competent person. That is, well-functioning sensory and cognitive skills will result in greater knowledge about what is there and its potential uses.

Second, the uses themselves will be more multifaceted. With greater personal competence, the functional and affective qualities of different environmental objects and contexts will be evident—people may choose, for example, to use or avoid an environmental object because it is affectively stimulating. Thus, in the end, the low-competence person's needs and preferences lead to a definition of the relevant environment that is relatively restricted, while the relevant environment is much richer and diverse for the high-competence person.

Another elaboration of the Lawton–Nahemow conception requires that environmental resources be differentiated into those that are given and externally generated and those that are created by the person. The distinction between those are by no means clear—in fact, both are obviously transactional, since any seemingly external environment has to some extent been chosen by the person, altered, or at the very least, been made legible or "effectable" (White, 1959) by the cognitive processes of the person. For convenience, however, such relatively stable features as social insititutions, social norms, community, neighborhood, dwelling-unit type, climate, transportation, or local amenities and services are often treated as fixed, while the proactively created environment is much more difficult to specify. In fact, the essence of a proactively created environment is its dynamic, temporally changing quality and its inability to be separated from the user. Examples of such transactional environmental resources are the cognitive map, the perceived or cognitively organized environment, the local amenities that are used, the state of maintenance of a home, the way it is furnished, the network of social relationships and the ways they function in relation to the subject, and so on.

The importance of the proactive stance has been supported by a tremendous amount of research activity on personal control and autonomous behavior in older people (Langer & Rodin, 1976; Schulz, 1976). Less ubiquitous has been research acknowledging that having personal control may or may not be the most successful or the most desired situation (Reid & Ziegler, 1981). For example, the perception of external control was shown to have favorable consequences in an environmental situation where degree of choice was very limited (Felton & Kahana, 1974). In general, however, evidence specifying the conditions under which internal or external control may be more favorable is lacking.

POSITIVE/NEGATIVE AFFECT

Another distinction relevant to the proactivity model is that between the outcomes of positive affect and negative affect. Long ago Bradburn (1969) demonstrated that under some conditions these two affect qualities are not opposites but in fact are unrelated. More recent research has continued to explore the relationship between these two types of affect, their antecedents, and their consequences (Diener, 1984; Warr, Barter, & Brownbridge, 1983; Watson & Tellegen, 1985). In an aged subject group, Lawton (1983) demonstrated that while negative affect was related primarily to physical health, positive affect was associated with contemporaneous, externally directed behavior, such as social activities and novel uses of time. An interesting line of research has shown that events passively responded to are likely to be associated with negative affect, while events sought out by the person may raise positive affect but not influence negative affect (Zautra & Reich, 1980).

General psychology has been concerned for some time with the range of experiences wished for and acted on by people, as well as what may be the consequences of stimulus variety versus sameness. Novelty (Fiske & Maddi, 1961), stimulus-seeking (Zuckerman, 1972), open versus closed ideologies (Rokeach, 1960), and other personal propensities have been studied essentially as traits. Throughout this period, adaptation level (Helson, 1964) has been invoked as a theoretical background for attempting to understand the temporal processes by which people's responses to stimulation change. Wohlwill (1974) elaborated these constructs, both theoretically and empirically, showing that changes in intensity of stimulation that deviate either positively or negatively from adaptation level may be judged as affectively positive.

The further development of the proactivity part of the model demands that we account better for the growth of competence and the enhancement of positive affect than our previous thinking about older people has done. It seems likely that personality factors such as augmentation–reduction, openness to experience, and affective self-regulation motivate people to search their environments for resources capable of maintaining a desired level of sensory, cognitive, and affective stimulation. Thus, an internal motivation leads the person to create a transactional environment of new press level. Those in full physical and psychological vigor will seek to increase some areas of competence. This proactive process is what Baltes, Dittman-Kohli, and Dixon (1984) have called "selective optimization with compensation," that is, the active selection of skills to enhance perceived and behavioral competence, together with relinquishing of skills that cannot be as effectively maintained.

Those with reduced levels of competence may well maintain an optimal overall level of functioning by reducing press level. It is possible to do so in a highly proactive manner. It may be a matter of limiting one's geographic range, of limiting one's exposure to emotionally activating relationships, or of rationing intellectual challenge. Nonetheless, to be one's own executive helps one remain in control. Self-chosen press reduction may enhance positive affect ("maximum comfort") or reduce negative affect. On the other hand, self-chosen press elevation or "resource promotion" has the capacity to elevate positive affect as described above. With poor judgment regarding one's own ability to handle the cognitive, affective, or behavioral load associated with increased press or resource level, however, an increase in negative affect may by risked.

It must be noted that the augmented model has a cognate for our initial assertion that there is no level of competence so low as to obviate the possibility of a congruent range of tolerable environmental press. Where proactivity is concerned, less formal research suggests that there is no level of competence so low that an increment in well-being cannot be attained by self-directed alteration of the environment.

An example of such proactivity has been described among extremely impaired older people who are enabled to remain living alone in the community by intensive in-home services (Lawton, 1985b). Their disabilities had forced them to restrict their behavioral space to the living room unless they had personal assistance in moving about the rest of the house. Given this impediment, most of them had constructed a new microenvironment that maximized the amount of knowledge and control they would have over the space they occupied most of the day. We called what they had created a "control center." It consisted of a well-chosen chair, oriented to afford a view of the front door and a maximum amount of the outdoors through a front window. The clock, telephone, television set, and a radio were handy. Tables or other surfaces were within arm's reach, tables littered with as many usable objects as possible, including photographs, mail, food, magazines, medicine, and so on.

Our view of a scheme of well-being suggests that although both docility and proactivity may have favorable outcomes, proactivity has an edge because it leads to shaping one's environment so as to maximize resource availability and use.

In conclusion, the self-constructed self is far from a logical impossibility. Research in aging demands that we look further to understand, first, how older people regulate their cognitive, affective, and behavioral lives. Research should also study how well the attainment of their preferred mix of stimulation, security, and affectively neutral, automatic-behavioral states leads to enhanced well-being.

REFERENCES

Baltes, P. B., Dittman-Kohli, F., & Dixon, R. A. (1984). New perspectives on the development of intelligence in adulthood. In P. B. Baltes & O. G. Brim (Eds.), *Life-span development and behavior* (Vol. 6, pp. 33–76). New York: Academic Press.

Birren, J. E. (1969). The aged in cities. *The Gerontologist, 9,* 163–169.

Birren, J. E. (1985). Age, competence, creativity, and wisdom. In R. N. Butler & H. P. Gleason (Eds.), *Productive aging* (pp. 29–36). New York: Springer Publishing Co.

Birren, J. E., & Hedlund, B. (in press). The metaphors of aging and the self-constructing individual. In J. Thornton (Ed.), *Metaphors of aging.* Vancouver, B.C.: University of British Columbia Press.

Bradburn, N. M. (1969). *The structure of psychological wellbeing.* Chicago: Aldine.

Carp, F. (1984). A complementary/congruence model of well-being or mental health for the community elderly. In I. Altman, M. P. Lawton, and J. W. Wohlwill (Eds.), *Human behavior and the environment: The elderly and the physical environment* (pp. 276–336). New York: Plenum Press.

Clayton, V. P., & Birren, J. E. (1980). The development of wisdom across the life-span. In P. B. Baltes & O. G. Brim (Eds.), *Life-span development and behavior* (Vol. 3, pp. 104–135). New York: Academic Press.

Culbertson, K. L., & Birren, J. E. (1973, November). *Mood and age—theoretical and methodological considerations.* Paper presented at the annual meeting of the Gerontological Society of America, Miami, FL.

Diener, E. (1984). Subjective well-being. *Psychological Bulletin, 95,* 542–575.

Felton, B., & Kahana, E. (1974). Adjustment and situationally-bound locus of control among institutionalized aged. *Journal of Gerontology, 29,* 295–301.

Fiske, D. W., & Maddi, S. (Eds.). (1961). *Functions of varied experience.* Homewood, IL: Dorsey Press.

Helson, H. (1964). *Adaptation level theory.* New York: Harper & Row.

Langer, E., & Rodin, J. (1976). The effects of choice and enhanced personal responsibility for the aged. *Journal of Personality and Social Psychology, 34,* 191–198.

Lawton, M. P. (1970). Ecology and aging. In L. A. Pastalan & D. H. Carson (Eds.), *Spatial behavior of older people* (pp. 40–67). Ann Arbor, MI: University of Michigan, Institute of Gerontology.

Lawton, M. P. (1976). The relative impact of congregate and traditional housing on elderly tenants. *The Gerontologist, 16,* 237–242.

Lawton, M. P. (1980). Environmental change: The older person as initiator and responder. In N. Datan & N. Lohmann (Eds.), *Transitions of Aging* (pp. 171–193). New York: Academic Press.

Lawton, M. P. (1982). Competence, environmental press, and the adaption of older people. In M. P. Lawton, P. G. Windley, & T. O. Byerts (Eds.), *Aging and the environment: Theoretical approaches* (pp. 33–59). New York: Springer Publishing Co.

Lawton, M. P. (1983). The dimensions of well-being. *Experimental Aging Research, 9,* 65–72.

Lawton, M. P. (1985a). Activities and leisure. In M. P. Lawton & G. Maddox (Eds.), *Annual Review of Gerontology and Geriatrics* (Vol. 5, pp. 127–164). New York: Springer Publishing Co.

Lawton, M. P. (1985b). The elderly in context: Perspectives from environmental psychology and gerontology. *Environment and Behavior, 17,* 501–519.

Lawton, M. P. (in press). Metaphors of environmental influences on aging. In J. E. Thornton (Ed.), *Metaphors of aging.* Vancouver, B.C.: University of British Columbia Press.

Lawton, M. P., & Cohen, J. (1974). The generality of housing impact on the well-being of older people. *Journal of Gerontology, 29,* 194–204.

Lawton, M. P., & Nahemow, L. (1973). Ecology and the aging process. In C. Eisdorfer and M. P. Lawton (Eds.), *Psychology of adult development and aging* (pp. 619–674). Washington, DC: American Psychological Association.

Lieberman, M. A., & Tobin, S. S. (1983). *The experience of age.* New York: Basic Books.

Neiswander, M. E., & Birren, J. E. (1973, November). *Love over the lifespan: A model and a methodological inquiry.* Paper presented at the annual meeting of the Gerontological Society of America, Miami, FL.

Reedy, M. N., & Birren, J. E. (1978, November). *How do lovers grow older together? Types of lovers and age.* Paper presented at the annual meeting of the Gerontological Society of America, Dallas, TX.

Reid, D. W., & Ziegler, M. (1981, November). *Longitudinal studies of desired control and adjustment among the elderly.* Paper presented at the annual meeting of the Gerontological Society of America, Toronto.

Rokeach, M. (1960). *The open and closed mind.* New York: Basic Books.

Schulz, R. (1976). Effects of control and predictability on the physical and psychological well-being of the institutionalized aged. *Journal of Personality and Social Psychology, 33,* 563–573.

Warr, P., Barter, J., & Brownbridge, G. (1983). On the independence of positive and negative affect. *Journal of Personality and Social Psychology, 44,* 644–651.

Watson, D., & Tellegen, A. (1985). Toward a consensual structure of mood. *Psychological Bulletin, 98,* 219–235.

White, R. W. (1959). Motivation reconsidered: The concept of competence. *Psychological Review, 66,* 297–333.

Wohlwill, J. F. (1974). Human adaptation to levels of environmental stimulation. *Human Ecology, 2,* 127–147.

Zautra, A., & Reich, J. (1980). Positive life events and reports of well-being. *American Journal of Community Psychology, 8,* 657–670.

Zuckerman, M. (1972). *Manual and research report for the Sensation-Seeking Scale.* Newark, DE: University of Delaware.

3

The Problem of Generations: Age Group Contrasts, Continuities, and Social Change

Vern L. Bengtson

The theme of generations is especially appropriate as we consider James Birren's influence on gerontology. Birren has been the seminal figure in the field, the father of more ideas and programs in gerontology than anyone else I know. Besides his own scholarly productivity (seen in over 250 publications to date), he has been a mentor and role model—a scholarly parent—to many gerontologists throughout the world. With the passage of years we have seen a second and third generation of scholars emanating from his research and teaching in a quite visible line of intellectual descent. And increasingly over time Birren has become paterfamilias with a host of projects and intellectual children demanding his attention and resources. Intergenerational continuity, both organizationally and personally, is, I think, one of the hallmarks of Birren's career. Intergenerational continuity and contrast is the theme of this chapter, with comments on parenting, generational innovation, and social change the focus.

Preparation of this chapter was supported by the National Institutes of Aging (#AG 04092). I am indebted to Robert Roberts for the data analyses reflected in this chapter and to Charlotte Dunham and Donna Polisar for assistance in literature review.

Generational relationships, especially parenting, have always been an uncertain enterprise within the human group. Parents invest prodigious amounts of time, energy, and material resources in the hope of producing offspring who will be happy, healthy, and wise. From my own research, it appears that one of the things parents strive for is continuity: indicators in the behavior of their children that they have achieved transmission of what is best in their own lives. And one of the things parents fear is discontinuity: evidence in the behavior of their children that they have chosen a very different set of values by which to live.

The problem of generations throughout recorded history reflects this tension between continuity and change, affirmation and innovation, in the human social order over time. On the one hand, the process of generational turnover and replacement implies change, both biological and social. Human offspring are not mere replicas of their parents; they represent subtly new genetic combinations in interaction with a unique environment (see Rossi, 1987). They also represent social innovation, as new participants in existing social systems (Braungart, 1984; Esler, 1972; Wood & Ng, 1980). Products of unique sociohistoric influences, young adults are the carriers of new perspectives and commitments which represent the potential for change as they encounter the existing social order (Elder, 1978; Ryder, 1965).

On the other hand, there is biosocial continuity across generations, seen in genetic and social similarities between parents and children. Despite change in both participants and social environment, the inertia of tradition does persist through the decades, as seen even in the rapidly changing world of science (see Cohen, 1985; Kuhn, 1962). Historical comparison suggests more stability through time in groups and societies, for better or for worse, than we often acknowledge (Allen, 1952). Social continuity is attempted most directly in the efforts of the older generation to transmit and preserve. Upholding a social order they have created, perhaps struggling to preserve positions of power, the ancien regime has attempted to enhance continuity of social order in the face of generational succession and innovation.

In this chapter the focus is on some consequences of the process of parenting, seen in the context of tensions between social continuity and innovation between generations. First, we will examine an instance of the macrosocial dilemmas of parenting, what future historians might look to as a modern example of the classic struggle between social generations—the protest movements of the 1960s, with their implications of generational rebellion and age-group cleavage. Second, we will explore some conceptual distinctions which help in understanding the generational causes of these movements: cohort, lineage, and period effects, mechanisms related to the succession of age groups whose operations are important in charting social change. Third, we will focus on issues of distinctiveness and influence in generational processes, reviewing evidence of value contrasts or similarities among youth, parents, and grandparents in the context of social change.

GENERATIONS, AGE GROUP CONTRASTS, AND THE DECADE OF PROTEST

The tension between generational continuity and change appears as a recurrent theme in writings concerning the social order, from Greek dramaturgy (see Datan, 1986) down to contemporary mass media (the *Time* magazine cover story of January 11, 1988 was titled: "1968—Like a Knife Blade, the Year Severed Past from Future"). The succession of one generation by another has often involved conflict between parents and children, youths and elders, as they attempt to negotiate a balance between continuity and innovation over time.

This tension, in sociological analysis, reflects two contrasting levels of social forces. The first concerns macrosocial manifestations of stability or change through time and the effects of generational turnover or cohort succession on broad patterns of social organization and culture. (The term *cohort* refers to individuals who share a common set of characteristics—in this case birth within a certain historical period—which causes individuals at similar ages to experience particular sociopolitical events at a common level of their biosocial development.)

At the societal level, social systems go through what Ryder (1965) described as "demographic metabolism" involving the entry of new cohort members and the departure of old. Population trends, economic indicators, and relationships among demographic groups and social positions are relevant variables here (Riley, 1985). One way to examine the macrosocial issue of cohort succession is as follows: Considering the birth rate, the dependency ratio, economic conditions, and political trends, are existing social structures and available roles appropriate to the population size and characteristics of birth cohorts growing up and moving into adulthood? If they are not, we have indicators of possible social change; the structures of existing society will be altered by the sheer size of a population moving into adult positions.

Cherlin (1981) refers to the potential "generation crunch" as social changes result from current demographic trends. The rapidity of change creates periods of transition in which role strain (e.g., conflicts between cohort-related roles of female, daughter, wife, mother, career person) may be experienced because people have not yet been socialized to the novelty of a particular structural change; they have not had enough "rehearsal" time in the negotiation of this life-course change. There is, of course, danger in a simplistic one-to-one juxtaposition of population processes and subsequent social change; but that cohort size influences macrosocial developments is clear (see Easterlin, 1980; Elder, 1984; Riley, 1985; Ryder, 1965).

The second level of analysis, the microstructural, focuses on the face-to-face negotiation of generational turnover and its social manifestations. The most obvious context is the family, a "unit of interacting personalities" (Burgess, 1926) composed of individuals at different levels in the unfolding succession of generations. Here the paradox of continuity versus change is most immediate, as family members negotiate norms, roles, and values consistent with their life-cycle positions. As Hagestad (1984) has pointed out, the negotiation of similarity versus individuation is complex, delicate, and unending—as well as frequently unnoticed. Outside the family (in microsocial encounters characterizing the workplace, the school, and voluntary organizations) interaction between "social generations" may be less affect-laden, but it still involves the negotiation of continuity or innovation between actors who have different "developmental stakes" in the maintenance of existing structures and patterns because of their differences in generational position (Bengtson & Kuypers, 1971).

The recent history of social movements in America provides a useful example of both macro- and microlevel phenomena involved in problems of parenting. More generally, they illustrate complex issues of generational succession and social change. Between 1960 and 1970 Americans, in company with most other Western industrialized societies, became concerned about political and social cleavages between age groups which appeared to be pulling apart the social order, and the "generation gap" suddenly became a theme of the era. Parenting, in the Benjamin Spock era, became viewed as a macrosocial problem (Leventman, 1982).

Five distinct but related social movements which had cohort and generation overtones characterized this "Decade of Protest."

The Civil Rights Movement

First was the emergence in 1959–1961 of protests for racial equality, in which predominantly youthful whites and blacks marched together to protest the discrepancy between values of equality and actions of organized discrimination. The Freedom Riders and their peaceful protests culminated in bloody confrontations which troubled an entire nation, suddenly aware that some of their youth were serious about social change. One legacy was the emergence of the New Left, with its innovative articulation of Marxist themes applied to old social inequities (Breines, 1982).

The Student Movement

The second started out more lightheartedly. The "free speech movement," begun in Berkeley in May 1964, was portrayed by the press as an attempt to shock the establishment by shouting obscenities over loudspeakers within the hallowed halls of academe (Nassi, 1981). But the issues were much more serious—college students questioning the legitimacy of their elders' complete control over educational governance, including the right to absolutely free speech. The Student Movement was born and quickly spread across the nation, causing changes in academic decision-making and instructional procedures.

Students appeared in classrooms with nonnegotiable demands regarding curricula; they "sat in" administrative offices to change in loco parentis rules. Academic administrators, authority figures representing unwanted parental control, were being challenged as never before in America (Miller & Gilmore, 1975). One explanation for such expressions of student unrest in the 1960s was in terms of generational conflict (Bettelheim, 1965), with college administrators the projection of parental control. Another explanation was sociodevelopmental; the protests were cohort expressions of age-related strivings for personal autonomy and social justice (Braungart, 1984; Feuer, 1969; Flacks, 1967; Whalen & Flacks, 1982).

The Anti-War Movement

At the same time, American military involvement in Southeast Asia was escalating. The Anti-War Movement, beginning in 1965–1966 and manned by predominantly youthful protesters in deadly earnest, drove a president from office and shook the views of America's elders concerning American foreign policy. Barbara Tuchman described these events in *The March of Folly: From Troy to Vietnam* (1984). Three issues, from her perspective, indicate a "march of folly" by a social generation: (1) policy adopted by the ruling cohort was perceived as counterproductive at the time, not merely in hindsight; (2) a feasible alternative course of action must have been available; and (3) the policy in question must have been that of a group or cohort, not merely an individual leader. For Tuchman, the Vietnam War involvement represents a classic case of the march of folly, especially in light of the increasingly vocal protests by those of the younger cohorts.

The Kent State Massacre presented the specter of law and order at any cost in the face of peaceful protest. By 1969 virtually every four-year college in the nation had experienced some form of organized protest on the part of youth questioning the moral and legal basis for an undeclared war directed by their elders (Wood & Ng, 1980). Most families experienced the disquieting effects of intergenerational confrontation about the war. It was a sobering spectacle of "America betraying herself" (Tuchman, 1984) in the eyes of many youths. The legacies of the Anti-War Movement are still potent factors in American politics (Surrey, 1982); a vocal minority of today's voters had their ideologies shaped by the rhetoric of that struggle (Jennings & Niemi, 1981; Wheeler, 1984).

The Women's Movement

The fourth major social movement of this decade began quietly in the 1960s but is the most obviously alive of any of the five today. Betty Friedan (1963) provided the manifesto for this movement, but even she did not foresee its momentum and scope (Friedan, 1984). Deckard (1983) has analyzed its political, sociological, and psychological origins; her conclusion is that while generational distinctiveness was not one of its original foci, college women of the 1960s have become its leading proponents. In order to create fundamental change, however, Deckard asserts that more than a decade of protest has been required. The Women's Movement has shown its strength and flexibility with the passage of time. Adaptation of tradition and consciousness raising has touched all groups (Freeman, 1975).

The Women's Movement is, in this context, perhaps the least indicative of generational cleavages. But it is also perhaps the 1960s movement least co-opted by other interest groups and subsequent political developments. In a recent examination of women students in business school, Cancian (1980) reviews the social and structural processes that contribute to the explanation and understanding of developments in the acceptance and endorsement of women in a male-dominated area of expertise. Continuing indicators of economic inequities between males and females suggest that the impetus to change is very much present as younger generations of women come of age.

The Counterculture Movement

A fifth example of social movements of the 1960s involved issues of life-style and personal values, more than issues of political or social inequity (Leventman, 1982; Yinger, 1982). Groups of youth began proclaiming values and behaviors which ran counter to the ethos of productivity, cleanliness, and bourgeois capitalism that appeared to characterize their parents' generation. Long hair, secondhand clothes, communal living, recreational use of drugs, casual sex—all reflected an emerging life-style in clear contrast to middle-aged, middle-class conventions of the 1960s. Their shock value was significant. Charles Reich (1970), in what became a rapid best-seller, called it "the greening of America."

Margaret Mead (1970) gave the movement a more scholarly and even more radical interpretation. She suggested that these youth were "immigrants in time" moving into a new cultural configuration and reversing prior mechanisms of socialization. Whereas in several hundreds of previous generations in human society children had learned from their parents, Mead asserted that social and technological innovation in the 1960s had accelerated so rapidly that in the next generation children would have to teach their parents how to survive, what is good or valuable, what is bad or maladaptive. The "counterculture" was for Mead the representation of a radical departure, and the older generation had better learn from it—and adapt. In one of the most memorable quotes of this period, Edgar Friedenberg (1969) gave the following judgment: "Young people today aren't rebelling against their parents; they're *abandoning* them" (p. 219). Middle-class parenting, from this perspective, had become irrelevant to the culture epoch that lay ahead.

COHORT, LINEAGE, HISTORICAL PERIOD

How can one account for such contrasts between cohorts of youth and elders, generations of parents and children, as seemed to surface in the 1960s Decade of Protest? What are the causes of social differences—and similarities—between cohorts and generations that surface in a particular historical period? Three concepts may be useful in examining the change and continuity in intergenerational comparisons evident between 1960 and 1970: cohort effects, lineage effects, and period effects (see Bengtson, Cutler, Mangen, & Marshall, 1985, for a more comprehensive discussion). Although all three effects are interrelated, each provides a slightly different perspective for viewing the succession of generations and subsequent social change. And each reflects the fact that parenting and its outcomes must be considered on three levels of change and development: individual development, family development, and historical development (Aldous, 1978; Elder, 1984; Haraven, 1977; Hagestad, 1984).

Cohort Effects

One explanation for contrasts—whether real or perceived—between individuals who differ in chronological age focuses on factors related to cohort experiences. In 1990 those who were born in 1965, compared with those born in 1925, will be products of very different sets of personal or sociopolitical concerns as well as life experiences. A first reason concerns individual development, seen in maturation and aging: Born in different points in time, the members of these two cohorts are at different points in the life cycle—at different stages psychologically, physiologically, and sociologically. Second, because they grew up at different points in historical time, they will also have experienced sociopolitical events differently, as those events were encountered at different stages of life-span development. The 1963–1972 war in Vietnam was undoubtedly experienced quite differently by most 20-year-olds compared to their 40-year-old parents, who were growing up during World War II and the global fight to save democratic government. These differences were reflected in public opinion data showing contrasts between cohorts in support for the Vietnam War (Cutler, 1976). There are good reasons for the existence of contrasts between individuals born at different points in time, either on the basis of maturation levels or cohort experience.

Karl Mannheim (1952) was the first modern sociologist to explicitly relate social change to age cohort effects. Because in the process of generational succession there is a "continuous emergence of new participants in the cultural process" (p. 293), he suggested that "each new generation comes to live within a specific, individually acquired framework of usable past experience, so that every new experience has its form and its place largely marked out for it in advance" (p. 296). It is because each new age cohort comes afresh upon the social scene and can see it with new perspective that new variations of old themes occur. It is from this "fresh contact" with existing social structures and values that a new weltanschauung, a new Spirit of the Age, evolves.

Mannheim (1952) also proposed the independent effect of what he termed the "generational unit"—members of a birth cohort who become "forerunners" in the pursuit of new alternatives to existing styles or causes. Not all people born at the same time share the same socialization or perceive historical events in the same way. "Only where contemporaries are in a position to participate as an integrated group in certain common experiences can we rightly speak of community of location of generation" (p. 198). Mannheim's notion of generation as a social unit is, however insightful, somewhat imprecise. While "the sociological phenomenon of generations is ultimately based on the biological rhythm of birth and death," his suggestion that we try "to understand the generation as a particular type of social location" (pp. 290–291) does not provide researchers with a clear conceptual definition upon which they might build.

Certainly no single characterization of youth in the 1960s would be completely descriptive of all members of that cohort. Rather, a range of styles could be discerned in the 1960s, termed revivalists, communalists, and freaks (Bengtson, Furlong, & Laufer, 1974). Although all of these youth were members of the same birth cohort, they were clearly not, in Mannheim's terms, of the same generational unit. Such other social structural variables as social class, race, and geographic location would be expected to influence intracohort differentiations.

Lineage Effects

Within the family there are also social forces which lead to contrasts and similarities between members who differ in chronological age. The "lineage gap" refers to real or perceived differences between generations within families, in the context of socialization forces to produce continuity. The family is a prototypical structure of social organization in which there is a series of statuses defined by ranked descent. Fathers and mothers, sons and daughters, grandchildren and grandparents, all form successive links in the flow of biological and social generations. This was the conceptual tool used by the very first recorded historians to put in order events of history: Writers of the Old Testament set off historical periods by lineage; in Egyptian records the time of a particular event was indexed by reference to the life span of a particular ancestor in the lineage chain.

Even though lineage descent can be an easily measurable phenomenon, it becomes complex in two ways. The first is suggested by Hagestad's (1981) aphorism that "generations do not file into families by cohort." That is, while many families have children who were born within one decade of historical time, others do not; and these cohort differences within one generation of family lineage may be quite important for within-family interaction. The second reflects the transfer from the microsocial level of a particular family to the macrosocial level of large aggregates such as the population of a nation. The term *generation* is very inexact in terms of social boundaries. We can refer to the generation of American college students today and assume, although we have used the term imprecisely, that at least the referent is to a group of individuals most of whom share a birthdate of 18 to 22 years earlier. But if we refer to the parental generation of the same students, we are faced with a group whose ages may be normally distributed, but the dispersion about the mean is quite large. This is because the timing of the births of children varies to a much greater extent than the timing of entry into college. This occurs for two reasons: the timing of first births varies widely within a population, and the spacing and number of children within a family interacts with parental age.

The emergence of four-, five-, and even six-generation families in 20th-century America suggests one aspect of the complexities of contemporary lineage relationships. Consider, for example, the consequences of "early grandmotherhood" and the intergenerational structure and dynamics of five generation families. Burton and Bengtson (1985) report a study of South-Central Los Angeles women who were as young as 27 when their daughters gave birth to a grandchild. For these women, the asynchronism between lineage generational placement (becoming a grandmother) and chronological age (less than 35 years) represented a major source of frustration regarding self-identity and satisfactions. In many of these families early pregnancy (at 11–14 years) had occurred in several generations. In some of the families, great-grandmothers and great-great-grandmothers were also interviewed; in one six-generation family the 91-year-old great-great-grandmother was living alone, while the young mother and infant were living with the 29-year-old grandmother. In this lineage, four of the six generations of women gave birth to their firstborn when they were between 11 and 14 years of age.

Are there inevitable differences between "generations" (or even "cohorts") by virtue of differential status as parent or child? Much of 20th-century psychology has assumed so, following the insights of Freud. The notion is that an inevitable and perhaps useful rebellion occurs as young children wish, first of all, to become their parents, and second, to take over the perquisite and power of parents. The theme of such generational conflict within families appears frequently throughout Western literature. One thinks, for example, of the aged King Lear at the end of Shakespeare's tragedy—an embittered king seeing his wishes disrespected, and crying, "Oh, the infamy that is to be a parent!"

But there is another and opposite dimension to the "lineage effect"—the processes of intergenerational socialization in which generations influence each other. Most often this is seen in transmission effects, from parents to children; but, as will be noted later, transmissions and influence are best seen as bidirectional (Acock & Bengtson, 1980; Glass, Bengtson, & Dunham, 1986; Hagestad, 1984). At any rate, the intended product of socialization is some degree of similarity in values and opinions, and the process involved is intergenerational interaction geared to enhancing similarity between parents and children.

Historical trends in family demography have created striking contrasts in intergenerational contact and interactions today compared with the last century (Treas & Bengtson, 1986). Declining fertility has effected differences simply in the number of individuals per generation within a given family. At the same time, more contact takes place today between generations geographically distant from each other, more than at any other time in American history, because of advances in telecommunications and travel technology that make such contact possible.

Whether the quality of intergenerational contact has changed as much as the quantity, over the last century, is not as certain. However, evidence from 19th-century American diaries and probate records suggest that disaffection between generations was often resolved by the younger generations moving west to the frontier and having little further contact with the older generations (Vinovskis, 1978).

Period Effects

Some sociopolitical events—wars, economic shifts, or political causes—are of sufficient magnitude and duration as to affect all groups within a society. These reflect *period effects*, and their analysis involves comparisons of perceptions, events, or attitudes at a given point in time to those of another time. An old Arab proverb suggests that "men resemble their times more than they do their fathers." It is to period effects that this proverb refers. When examining relationships between generations or between age groups within society, it is instructive to compare them to such relationships in other periods of history.

Almost two decades later, pronouncements like Friedenberg's (1969) and Mead's (1970) quoted above, concerning radically new generational or age-cohort configurations in our culture, appear extreme and outdated. On contemporary college campuses there are few reminders that the 1960s clash between generations was so pervasive and portentous (Breines, 1982). Age groups in the 1980s do not seem to oppose each other much; rather, as in the 1950s, members of differing cohorts appear to espouse similar concerns about the economy, taxes, jobs, and America's foreign policy. The 1958 of Eisenhower, not the 1984 of Orwell, seems the more appropriate analog to America in 1988. Perhaps the apparent swing of the historical pendulum is exactly the point: that both change and continuity are exhibited by social groups composed of different generations, at contrasting points of historical time.

Period effects may not, however, be as unique as they may appear at the time. For example, laments of older generations concerning the young can be seen throughout the philosophical and moral literature of Western civilization. Plato's observations about the young of his era being undisciplined and unmotivated has been repeated as solace for those in the middle generation who may find that their own children's lack of industry fails to meet their ideals. It may be that the so-called generation gap is really not more serious, even in today's fast-changing society, than it was in earlier historical periods. It may be that there is an inevitable "period gap," which only seems to reflect generational cleavages (the youth of any era being the most outspoken advocates of new interpretations or social processes and events). This may be interpreted in statistical terms as a cohort x period interaction.

Applications to the Decade of Protest

In short, when we examine intergenerational interaction and the differences which appear between generations, it is necessary to distinguish among at least three causes and their potential interactions. Each effect—cohort, lineage, and period—may create natural and inevitable differences in attitudes, values, and behaviors of the individuals involved in intergenerational interaction. But each also can be seen as effecting some degree of continuity. Maturational changes can be expected to bring children closer to their parents' orientations as they mature in adulthood; lineage effects involve transmission between generations and therefore greater similarity; period effects involve all contemporaneous generations experiencing similar sociohistorical events.

What do these perspectives suggest about the rise of social movements in the 1960s and their roots in generational cleavage? Was this a classic case of generational rebellion, and if not, what were its more likely antecedents?

Certainly cohort effect explanations are the most plausible. At the macrosocial level the antecedents of these social movements can surely be traced to one factor involving the succession of generations: population processes which resulted in a "baby boom" coming of age. The sheer numbers of youthful recruits into the American social order of the 1960s could have been expected to change existing configurations of both roles and ideologies. That few commentators had noted this or predicted the massive social changes that resulted is perhaps astonishing—especially in the wake of classroom building and school expanding that occurred in the 1950s as these "baby-boomers" went off to elementary school. But Americans are notoriously optimistic, and perhaps opimism breeds conservatism in terms of the belief that things will pretty much stay the same or get better. Certainly we expect that our children will have a better life than we had—a hope that, in today's economic climate, is perhaps unrealistic. That the protesters of the 1960s are today mostly middle-class producers and consumers, worrying about mortgages and financing their children's college educations, does not deny that their sociopolitical ideology was affected by their cohort experiences (see Fendrich, 1974; Nassi & Abramowitz, 1976; Wheeler, 1984).

The interaction between cohort and period effects is also a plausible explanation of these social movements, reflecting their contrasting meanings for contemporaneous age groups. The older cohorts had experienced World War II directly (a cohort effect); this was certainly important in their interpretation of the Vietnam War and the student protest movement (period effects). This created a tension vis-a-vis younger cohorts, who had not experienced World War II but who were impacted by the 1960s protest movements (a period effect most affecting their cohort).

At the microsocial level the predictors of the five social movements are less clear. Certainly there was evidence of contrasts in values and sociopolitical attitudes between social movement participants and their elders. However, having and espousing different values is not necessarily a rejection of one's parents, nor is it an invalidation of the transmission of values via one's parents. Deutscher (1973) has noted that Americans are socialized to believe from early childhood that change is both inevitable and good. If Americans do have a national ethos that applauds and internalizes progress (i.e., change), as Deutscher asserts, familial continuity may merely be expressed in different forms (Nassi, 1981). With the passage of time, sociohistorical events tailor these changing forms. Yet, in speaking of changing configurations in the life course, Uhlenberg (1978) cautions that it is important not to ignore the remarkable stability *and* adaptability of the American family. Temporal dimensions create shifts in demographic and social aspects of the family life course. These subsequent shifts give rise to questions and future challenges in parenting, grandparenting, and intergenerational continuity.

Moreover, it does not appear to be the case that the young radicals of the 1960s were "rebelling" against their parents (Flacks, 1967). On the contrary, the left-wing college students were, as a group, more in political agreement with their parents—liberals themselves—than were the less active, more conservative students (Flacks, 1967; Nassi, 1981; Whalen & Flacks, 1982; Wood & Ng, 1980). A similar conclusion was drawn by Keniston (1968) in his analysis of the psychosocial development of "young radicals." These studies suggest that the youthful protesters in the forefront of the 1960s social movements were implicitly carrying out an agenda of intergenerational continuity, not change, and applying parental pronouncements of involvement, consistency, and principled living far more extensively than their parents had foreseen.

Their actions undoubtedly influenced their parents—and their grandparents. Nowhere is this more clear than in the adoption by middle-class, middle-aged Americans of the paraphernalia of counterculture style: Levis and beards and bell-bottomed pants, not to mention increased use of psychotropic drugs and a greater openness in exploring "the joys of sex" (Comfort, 1968). In the political realm, the influence of youth on their elders is less clear, since the issues of age cleavage quickly became co-opted by other sociopolitical issues—Watergate, the energy crisis, inflation—in a classic example of period effects muting the cohort or lineage contrasts evident a few years earlier.

DISTINCTIVENESS AND INFLUENCE

Two assumptions can be seen in analyses concerning parent-child interaction prior to the Decade of Protest (as discussed in Troll and Bengtson, 1979). The first concerns the inevitability of substantial differences, if not conflict, between generations (Bettelheim, 1965; Davis, 1940; Mannheim, 1952). As new cohorts of youth become independent adults, they attempt to maximize their distinctiveness from the parental generation. The second assumption concerns the direction of influence in socialization. Most studies that have examined similarity between youth and parents adopt a unidirectional model of transmission, in which influence is seen as passing down the generations from parent to child, from elder to youth. This assumption is based on another premise: that individuals' potential for change and development is highest at the beginning of life and is less after adolescence.

However, recent evaluations of intergenerational interaction (see Acock, 1984; Glass et al., 1986; Hagestad, 1981) are based on an alternative assumption, that the interactions involved in socialization affect all participants, whatever their time of life. This position leads to two corollaries. The first is that distinctiveness between parents and youth is modified by processes enhancing solidarity between generations; the second, that socialization is bilateral, not unidirectional. If youth adopt distinctive behaviors and values, for example, it is possible that these will influence and perhaps modify the prior orientations of their parents. From this perspective, each interacting generation will change, develop, or be socialized anew in the ongoing process of negotiating generational emergence. This logic is consistent with the emerging life-course perspective within the sociology of aging (Braungart & Braungart, 1986; Dannefer, 1984; Featherman, 1982; Featherman & Lerner, 1985).

Socialization and Transmission

Primary socialization can best be seen as a process of ongoing negotiation between generations, in which the younger member takes on new age-related roles and the older member provides information, normative guidance, and support. However, it is important to note that relations between parents and children at each stage of the life course reflect influence processes that are both reciprocal and continuously changing. The patterns of such influence within families are altered, directly or indirectly, by societal processes that themselves can be viewed in generational terms and that derive from period and cohort themes, on the one hand, and family themes on the other.

For example, in the process of transition into adulthood, a new cohort of youth, or at least what can be called (following Mannheim, 1952) a band of "forerunners" in that cohort, may strike a unique theme or keynote that sets itself off from its parents and elders—one explanation of what happened in the 1960s. But as the research by Flacks (1967) and others suggests, such a keynote often derives from the salient leitmotifs of the forerunners' families. From this perspective the power and persistence of the new themes depend upon their congruence with ongoing historical processes within the culture, as well as within the families involved (Troll & Bengtson, 1979).

Transmission has the connotation of sequentially passing on information in a linear fashion from one unit of a system to another (such as generations within a family). But it is important to note that transmission also implies exchange—that the actions of each unit in the sequence are influenced by the actions of the others (Dowd, 1980). In short, there is feedback among elements of the socialization system (Hernes, 1976). Under such conditions sequence or causal ordering may be difficult to ascertain.

This is the case when we attempt to answer questions concerning intergenerational transmission or contrast in the 1960s. Indications of similarities or differences between parents and children compared at one point in time can be used to examine three issues involved in generational analysis (Bengtson et al., 1974). The first involves descriptions concerning the degree of similarity or difference between age cohorts or generations. To what extent do members of different generations appear to be distinctive from or to replicate each other, in behaviors, attitudes, and orientations?

A second issue concerns the causes of cohort and generational contrast or similarity. Can differences between generations be attributed to contrasts in developmental or ontogenetic status, or are they better traced to cohort effects—to being born and coming of age at different points of history, evidencing the differential influence of sociohistorical trends (Elder, 1978)?

The third issue involves the sequence of socialization influence. Is it at all possible to infer "transmission" from evidence for "similarity"? If so, who influences whom—do parents not learn from youth, as well as the reverse (Glass et al., 1986)?

Distinctiveness

Two theoretical extremes are suggested by the scholarly literature to date concerning similarity or contrast between age groups (reviewed in Bengtson, 1970; for an earlier view see Davis, 1940). One emphasizes the inevitability of differences because of different locations in developmental and historical time. Each cohort must deal anew with issues of identity, intimacy, values, and appropriate behaviors as it moves into adulthood and comes into "fresh contact" with established configurations of culture. This is especially true in periods of rapid social change (Davis, 1940; Reich, 1970). Mead (1970) argued that in the "prefigurative" culture which she felt was emerging the old must learn from the young, since the pace of contemporary technosocial change has become so rapid.

The opposite position minimizes generational contrasts. Apparent differences between generations are temporary; children differ from their parents primarily because of ontogenetic developmental status (Adelson, 1970; Davis, 1940), and youth differ from their elders because of the necessity to develop their own identity (Bettelheim, 1965). Adolescents are different from older adults, but when youth in turn become middle-aged or old, they will then resemble their parents and grandparents. Media portrayals of today's "Yuppies" as a birth cohort emerging to espouse conservative values after they outgrew their youthful indiscretions (involving social protests during the 1960s) is, perhaps, a good example of this perspective.

An accurate picture of generational distinctiveness probably lies between these two extremes. But relevant to both extremes may be the "generational stake" each cohort has in maximizing or minimizing its perception of continuity (Bengtson & Kuypers, 1971). On the one hand, parents may wish to minimize their offspring's distinctiveness. The perspective of middle-aged parents on the next generation is in part a product of their own life-span status: Erickson (1950) has termed this "generativity," suggesting that one theme of middle age is building continuity. The effort and commitment parents have invested in raising their children, their present diminished influence on their children, and the recognition of their own mortality make it important that the next generation "carry on." On the other hand, their children, establishing what Erikson calls "identity distinctiveness," are looking forward to an independently constructed life ahead. Needing to express their uniqueness, they view their parents' goals from a different perspective and may minimize apparent continuities with their parents and grandparents. Each generation thus has a different stake in intergenerational continuity or distinctiveness. (See discussions of this point in Lerner, 1975; Lerner & Knapp, 1975; Thompson, Clark, & Gunn, 1985.)

Values, Generations, and Change between Historical Periods

Contrasts in value orientations have frequently been suggested as a marker of the interplay between cohort distinctiveness and social change over time. If younger cohorts evidence substantial contrasts in values compared with older cohorts, this can be interpreted either as a prediction that other social change will follow or a confirmation that important changes have already taken place. Few empirical analyses are available to chart such contrasts; what is required are longitudinal or time-sequential observations in a sample reflecting several birth cohorts and/or lineage generation of respondents.

Data from the USC Longitudinal Study of Generations provides at least limited assessment of this issue. The study was begun in 1971, shortly after the events of the Decade of Protest reviewed above; these events suggested the hypothesis of marked age-group contrasts in basic value orientations. In the Time 1 assessment 2,044 individuals—grandparents, parents, and middle-aged children sampled from a Southern California Health Maintenance Organization—provided data ranking, in order of importance, of 16 items reflecting values in four domains: humanistic, materialistic, individualistic, and collectivistic (for details see Bengtson, 1975).

The results from the 1971 survey revealed some support for the expected contrasts between the generational groups surveyed (Bengtson & Lovejoy, 1973). It was anticipated that individualistic values (personal freedom, an exciting life) would be ranked more highly by the youth, and this was the case (see Figure 3.1). Similarly, it was expected that collectivistic values would be most highly endorsed by the grandparents: patriotism and religious participation were considered more relevant values to this age group than to their children or grandchildren. This also was supported by the data.

However, there were some unexpected patterns in the comparisons between generational groups, of which several were in sharp contrast to contemporary wisdom regarding cohort- and generation-related patterns in value orientations. On humanistic values (for example, the item "a world at peace") it was expected that youth would evidence the highest endorsement; after all, it was the youth who appeared in the forefront of peace demonstrations in the 1960s. However, of the three age groups, it was the grandparents who ranked this value item the

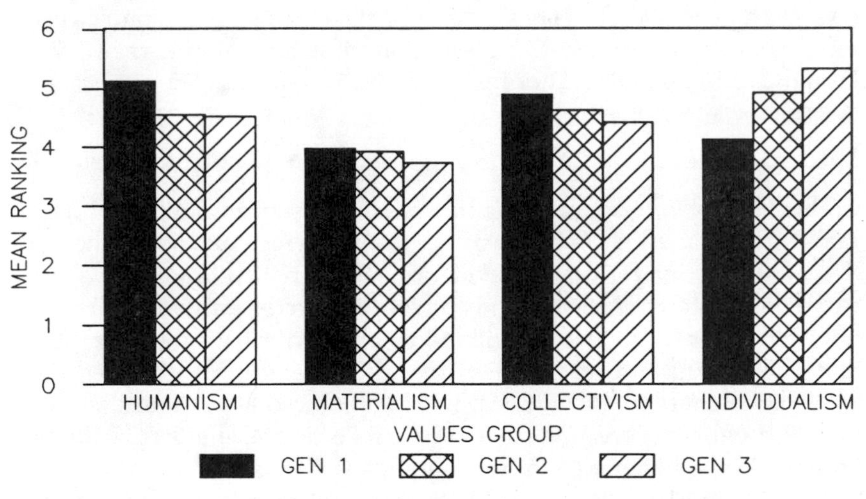

F Values Corresponding to Between-Generation Values Ranking ANOVAs by Values Grouping: 1971

1971	F	DF	P
Humanism	9.35	1025,2	<.001
Materialism	2.92	1025,2	N.S.
Collectivism	10.32	1025,2	<.001
Individualism	62.49	1025,2	<.001

FIGURE 3.1a Mean rankings by generation and values group, 1971.

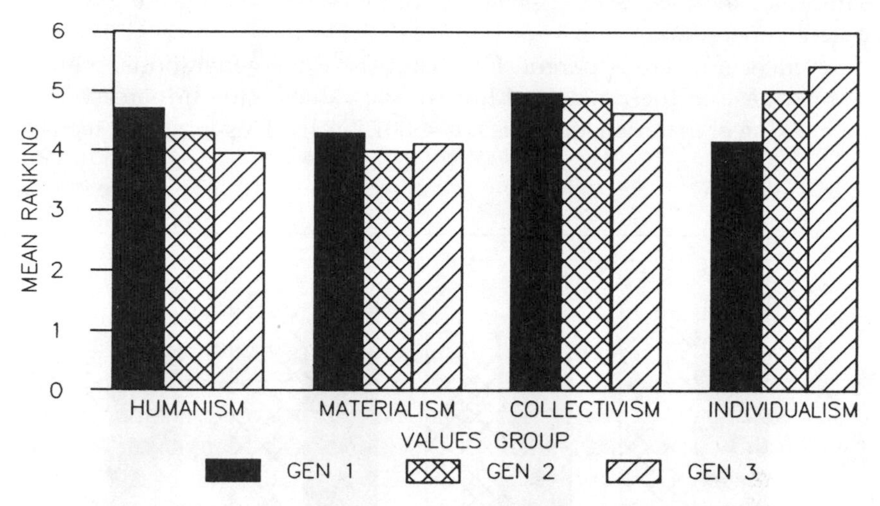

F Values Corresponding to Between-Generation Values Ranking ANOVAs by Values Grouping: 1985

1985	F	DF	P
Humanism	25.06	1025,2	<.001
Materialism	6.03	1025,2	<.01
Collectivism	6.77	1025,2	<.01
Individualism	87.56	1025,2	<.001

FIGURE 3.1b Mean rankings by generation and values group, 1985.

highest. Similar patterns appeared on "service to mankind" and "equality of mankind"—each valued most highly by the grandparents, second by grandchildren, third by the middle-aged parents. These rankings on humanistic value orientations suggested that common perceptions about generational differences were both stereotypic and inaccurate.

Over time, how might such value patterns change? Do the age-group patterns evident in 1971 persist over the years, suggesting some degree of cohort distinctiveness; or do they indicate age-related (maturational) contrasts? Is there evidence of period effects, resulting in age groups changing together?

Data from the 1985 (Time 2) survey assessment of these same family members provide some indications of persistence and change in value orientations. Figure 3.2 indicates contrasts between the two time periods in the four dimensions of value orientations, for each of the three generational groups. The sample size is 1,028, reflecting only Time 1 and Time 2 participants who furnished complete data. There was considerable sample attrition since 1971, especially due to death among the grandparent generation.

Several trends are apparent. First, over the three generations, there is a decrease in the ranking of humanistic values and an increase in the ranking of materialistic values ($p < .001$, t-test). This implies a secu-

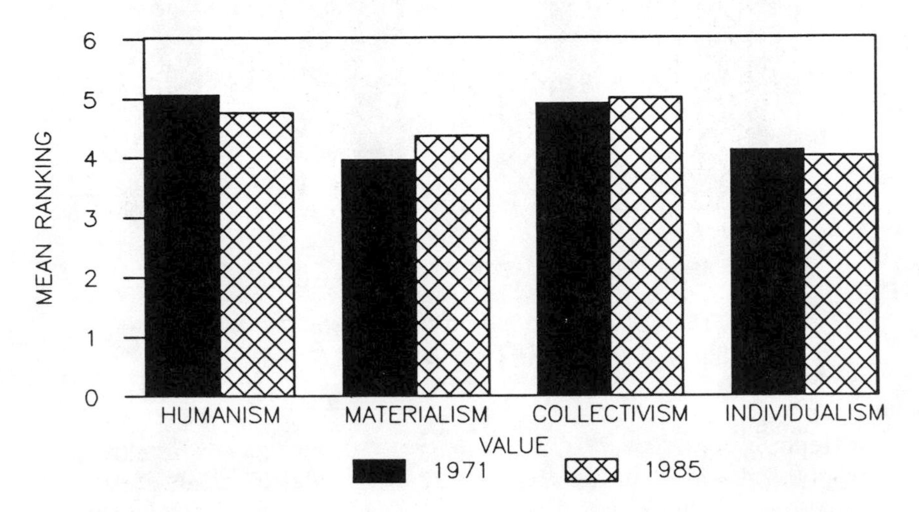

FIGURE 3.2a Generation one mean value rankings, 1971 versus 1985.

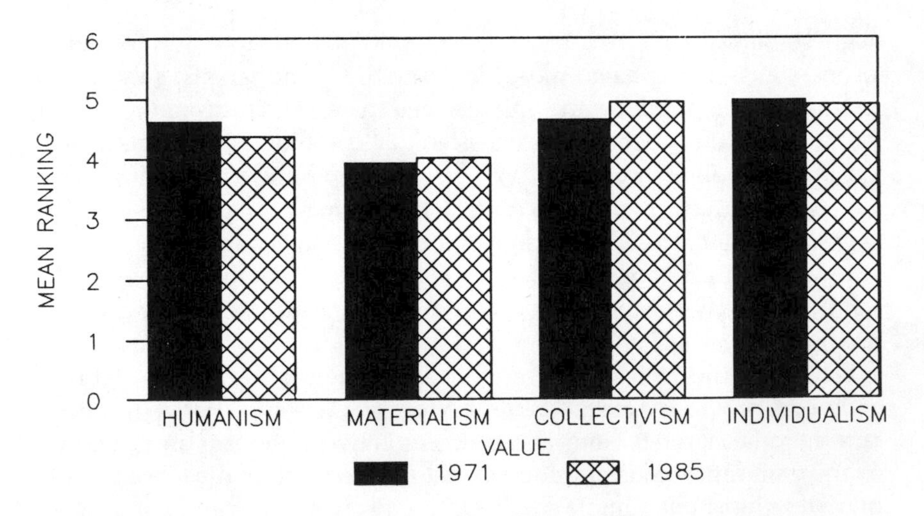

FIGURE 3.2b Generation two mean value rankings, 1971 versus 1985.

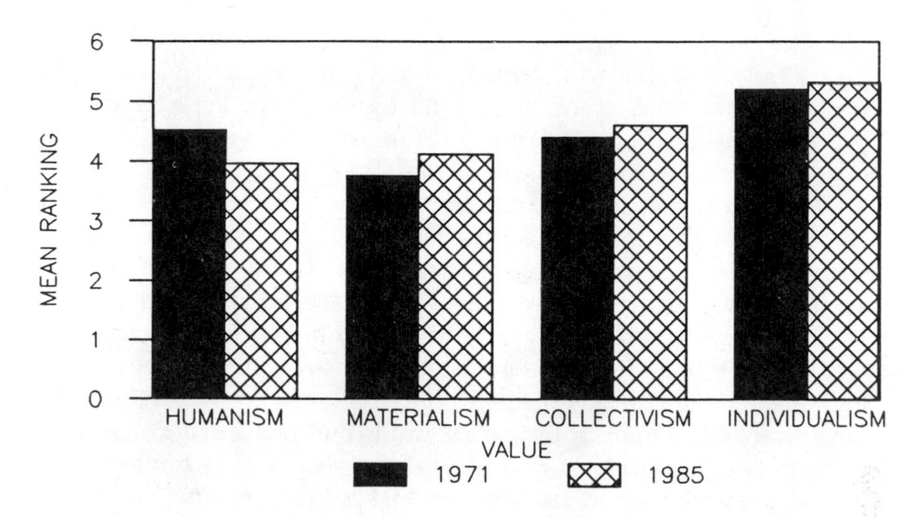

FIGURE 3.2c Generation three mean value rankings, 1971 versus 1985.

lar trend, a period effect, mirroring what has been suggested regarding political and ideological patterns in America of the 1980s: movement away from priorities reflecting "a world at peace" and "equality of mankind" and toward "financial comfort" as well as "recognition for personal accomplishments" (specific items reflecting these two value dimensions).

Second, these shifts are most apparent within the younger generation of respondents. Mean endorsement of humanistic values dropped most among the grandchildren (average age, 19 in 1971; 33 in 1985), second most among the grandparents (average age 79 in 1985), and least among the middle-aged parents. The same pattern is observed with regard to increases in materialistic values. Of the three generations, the middle changed the least.

These data, while only preliminary, suggest relatively little support for a *cohort* hypothesis of generational distinctiveness in values. They do indicate partial support for a contrasting hypothesis, that of *maturational* change over time. Most consistent with the data are *period effect* explanations. One other model, involving *lineage* effects, may be important in explaining these findings: Because the age cohorts surveyed are members of three-generation families, reciprocal influence mechanisms may be substantiated in future analysis. It is to the issue of reciprocal influence between generations that we turn next.

Influence

Theories regarding social influence are also important in understanding contrasts and similarities among cohorts and generations. As noted earlier, recent models of socialization have emphasized the bidirectional process of negotiating similarity and difference between parents and children (Smith, 1983; Glass et al., 1986). For example, Hagestad (1977) asked mothers of college-age children whether they felt their children had tried to influence them during the past 2 or 3 years and whether such efforts had been successful. Three-fourths of the sample recalled such attempts, and about two-thirds of these reported them to have had an effect. A number suggested that without such influence they would have found the events of the late 1960s to be much more foreign and threatening. One volunteered, "You think that boys with long hair are a strange and dangerous species until your own son becomes one of them and you discover that he is still the same kid—honest, concerned about the world around him, not wanting to hurt anybody" (Hagestad, 1977, p. 19). Thus, perceptions of distinctiveness are not incompatible with perceptions of bidirectional influence.

The issue of perception—of attribution to the other—is also important in exploring apparent generational dissimilarities as well as influence. Several studies (Acock & Bengtson, 1980; Bengtson & Kuypers, 1971; Gallagher, 1976; Lerner, 1975) demonstrate the difference between actual attitudes or values expressed by youth and parent respondents, on the one hand, and the attitudes and values each perceives the other to have, on the other. Each generation misattributes, to some degree, the attitudes of the other, although in opposite directions. Late-adolescent children exaggerate the differences between their own attitudes and those of their parents, while their parents minimize this difference (Bengtson & Black, 1973; Thompson & Walker, 1984). The actual parent-child contrasts fall between these two extremes. The discrepancy between actual and perceived orientations may indeed enter into the negotiation of differences involved in cross-generational influence during primary socialization and into adulthood.

The contrast between actual versus perceived differences takes on importance when one examines the outcome of parental influence attempts. For example, in the socialization of attitudes, which is the better predictor of youth's orientations: what parents actually think (their stated attitudes) or what their children think they think (perceived attitudes which their children attribute to them)?

One line of theory, that of behaviorists, suggests that it is the *actual* opinions or goals of parents which constitute the model of influence;

through another sequence of rewards or punishments, orientations expressed by children which the parents consider valuable are explicitly reinforced. By contrast, another theoretical perspective—represented by cognitive psychologists and the sociological sector of symbolic interactionists—would suggest that it is the orientations *attributed* to the parents by the children which are more influential.

In one examination of this issue from the USC generations project (Acock & Bengtson, 1980), the stated attitudes (self-respect responses reflecting "your own opinion") of mothers, fathers, and youths from 466 family triads were examined on political and religious questions. The young adult children were also asked to predict their mothers' and fathers' responses (attributed attitudes) to these issues. The research question involved the path of influence in the "social construction of reality" evidenced between generations. By exploring the actual versus perceived attitudes of similarity between parents and youth, an attempt was made to resolve contradictions between claims of similarity and claims of differences (or distinctiveness) between generations. The attitude items touched upon issues of work, government, law and order, business, college demonstrators, civil rights, welfare, marijuana, and religion.

Results indicated several findings. First, the most consistent evidence was that the "generation gap" was far more apparent in the minds of the children than was evident in contrasts between youths' and parents' actual reports of their sociopolitical attitudes. There was a persistent misperception on the part of the children, based perhaps on the assumption of an age-related polarization between children and their parents; the result was discrepancy between perceived (attributed) attitudes and actual opinions of parents.

Sherif, Sherif, and Nebergall (1965) noted that an individual's opinion in one situation often becomes an "anchor" in the placement of other opinions. Perhaps, in their children's attributions, parent opinions in one area become "anchored" opinions in other areas; if so, children begin to see disagreement as polarization and agreement as assimilation. Second, misattribution may be related to generational contrasts in ontogenetic development, reflecting the "generational stake" hypothesis mentioned earlier (Bengtson & Kuypers, 1971). Parents have greater levels of investment in the long-term career of the parent-child relationship, which causes them to see more continuity and cohesiveness in the relationship than the child may perceive. Third, misattribution may be due to lack of communication between generations or even deliberate misrepresentation as reflected in the parental adage "Do what I say, not what I do."

A second finding was that the *attributed* parental attitudes (what children perceived to be their parents' attitudes) were much more predictive of the children's *own* attitudes than were the *actual* opinions currently stated by the parents. Children also perceived strong parental consensus between mothers and fathers, while actual similarity between their attitudes was slight. These cohort level contrasts can be describe as "polarized misattribution"—not only did children inaccurately perceive their parents' opinions, but they saw them as more extreme (in the conservative direction) than they actually were.

The major implication of this study is that children are strongly influenced by parental attitudes, but only as these attitudes are perceived (attributed), not as they actually are. If children have perceived and thereby construct a "generation gap between themselves and their parents," it is important to recognize the social construction of these differences and to pursue its source. The distinction between actual and perceived attitudes is crucial in attempting to explain contradictory outcomes. It is also essential for socialization theorists to pay closer attention to attribution processes when charting intergenerational processes of influences.

CONCLUSIONS

The problem of generations, from mankind's earliest writings down to contemporary mass media accounts, involves the tension between continuity and change among age groups in the succession of social order. How does generational succession relate to the balance between change and innovation in human societies? What are the effects of maturation (age), historical placement (cohort), and emerging sociohistorical events (period) on generational differences and continuities?

Social scientists have recently been examining this classic problem of generations from a variety of perspectives, involving life-course, sociohistorical, and familial parameters of contrasts between age groups. This chapter has reviewed a variety of studies and perspectives regarding generations, continuities, and social change, with a focus on the social movements of the 1960s, which appeared to reflect a dramatic discontinuity between youth and their elders.

Four conclusions can be drawn reflecting the data reviewed in this chapter regarding generations, age group contrasts, and social change. First, conceptual distinctions among cohort, maturation, period, and lineage effects can account for many of the apparent contrasts between individuals of differing age groups at any point in time. Cohort effects indicate factors related to real or apparent differences between individuals born at different points in historical time; they are not to be confused with maturation effects, reflecting consequences of individual aging processes. Lineage effects represent the bidirectional nature of intergenerational socialization, which can lead to continuities despite cohort and maturation differences. Period effects can be seen in the impact of sociohistorical events as they impact on contemporaneous age groups.

Second, the most plausible explanations for the youth-led protest movement of the 1960s involve interactions among these four effects. The sheer size of the cohort coming into postadolescence during this decade, compared to earlier cohorts, coupled with new historical events such as the war in Vietnam and the civil rights protests, created a macrosocial environment where change was valued, and came to be demanded, by many youth. The experiences of their elders—coming of age during the Great Depression and World Wars I and II—was greatly different, leading to the appearance of generational discontinuity. However, lineage effects were also evident; many youthful protesters during this decade appeared to be quite similar to their parents in orientations and concerns. And indeed, the "generation gap" cleaveages which were so much a focus of the mass media during the 1960s seem not to be sustained over time; period effects appear to have more enduring consequences in attitudes and value orientations.

Third, issues of distinctiveness among emerging cohorts appear to be mitigated by countervailing processes of intergenerational influence, which are bidirectional: parents being influenced by their offspring, as well as the reverse, with both changing as the result. This process of ongoing negotiation between generations is only beginning to be explored in recent research, but is provides an important alternative to traditional theories of socialization and intergenerational transmission.

Fourth, the role of perceptions—the phenomonology of attributed contrasts or continuities—is important in assessing differences between age groups. There are good reasons why the members of differing generations may distort the orientations of each other (in the service of generational continuity or distinctiveness). But these can be apparent, not real, differences, and ongoing interaction can bring to each generational participant a more realistic assessment of the other's views.

It may be, as folk wisdom through time has insisted, that history repeats itself. Nevertheless, different generations seem to feel they have their own distinctive historical imprint, even if they experience similar frustrations in maturation and encounter similar challenges. Each emerging cohort may be convinced that it produces a unique content and form; each cohort may also believe that the past generations cannot understand what they are experiencing. This perspective of distinctiveness may be, paradoxically, one of the traditions that is maintained over time in the context of succeeding generations.

REFERENCES

Acock, A. C. (1984). Parents and their children: The study of intergenerational difference. *Sociology and Social Research, 69,* 2–22.

Acock, A. C., & Bengtson, V. L. (1980). Socialization and attribution processes: Actual versus perceived similarity among parents and youths. *Journal of Marriage and the Family, 42,* 501–515.

Adelson, J. (1970, January 18). What generation gap? *New York Times Magazine,* p. 10ff.

Aldous, J. (1978). *Family careers: Developmental changes in families.* New York: Wiley.

Allen, F. (1952). *The big change: America transforms itself, 1900–1950.* New York: Harper & Row.

Bengtson, V. L. (1970). The "generation gap": A review and typology of social-psychological perspectives. *Youth and Society, 2* (1), 7–32.

Bengtson, V. L. (1975). Generation and family effects, in value socialization. *American Sociological Review, 40,* 358–371.

Bengtson, V. L., & Black, K. D. (1973). Intergenerational relations and continuity in socialization. In P. Baltes & K. W. Schaie (Eds.), *Life-span developmental psychology and socialization* (pp. 207–234). New York: Academic Press.

Bengtson, V. L., Cutler, N. E., Mangen, D. J., & Marshall, V. W. (1985). Generations, cohorts, and relations between age groups. In R. Binstock & E. Shanas (Eds.), *Handbook of aging and the social sciences* (pp. 304–338). New York: Van Nostrand Reinhold.

Bengtson, V. L., Furlong, M. J., & Laufer, R. S. (1974). Time, aging, and the continuity of the sociological structures: Themes and issues in generational analysis. *Journal of Social Issues, 30*(2), 1–30.

Bengtson, V. L., & Kuypers, J. A. (1971). Generational differences and the "developmental stake." *Aging and Human Development, 2,* 249–260.

Bengtson, V. L., & Lovejoy, M. C. (1973). Values, personality, and social structure: An intergenerational analysis. *American Behavioral Scientist, 10,* 880–912.

Bettelheim, B. (1965). The problem of generations. In E. Erikson (Ed.), *The challenge of youth* (pp. 76–109). New York: Anchor.

Braungart, R. G. (1984). Historical generations and youth movements: A theoretical perspective. *Research in Social Movements, Conflict and Change, 6,* 95–142. Greenwich, CT: JAI Press.

Braungart, R. G., & Braungart, M. M. (1986). Life course and generational politics. *Annual Review of Sociology, 12,* 205–231.

Breines, W. (1982). *Community and organization in the new life: 1962–1968.* New York: Praeger.
Burgess, E. W. (1926). The family as a unit of interacting personalities. *Family, 7,* 3–9.
Burton, L. C., & Bengtson, V. L. (1985). Black grandmothers: Issues of timing and continuity in roles. In V. Bengtson & J. Robertson (Eds.), *Grandparenthood* (pp. 304–338). Beverly Hill, CA: Sage Publications.
Cancian, F. M. (1980). Rapid social change: Women students in business schools. *Sociology and Social Research, 66*(2), 169–183.
Cherlin, A. (1981). A sense of history: Recent research on aging and the family. In B. Hess & K. Bond (Eds.), *Leading edges* (pp. 21–50). Washington, DC: National Institutes of Health.
Cohen, J. B. (1985). *Revolution in science.* Cambridge, MA: Harvard University Press.
Comfort, A. (1968). *The joys of sex.* New York: Random House.
Cutler, N. E. (1976). Generational analysis and political socialization. In S. A. Renshon (Ed.), *Handbook of political socialization: Theory and research* (pp. 321–346). New York: Free Press.
Dannefer, D. (1984). Adult development and social theory: A paradigmatic reappraisal. *American Sociological Review, 49,* 100–116.
Datan, N. (1986). Oedipal conflict, platonic love: Centrifugal forces in intergenerational relations. In N. Datan, A. Green, & H. Reese (Eds.), *Life span developmental psychology: Intergenerational networks* (pp. 29–50). Hillside, NJ: Erlbaum.
Davis, K. (1940). The sociology of parent-youth conflict. *American Sociological Review, 5,* 523–534.
Deckard, B. S. (1983). *The Women's Movement.* New York: Harper & Row.
Deutscher, I. (1973). *American families at the crossroads.* New York: Wiley.
Dowd, J. J. (1980). *Stratification among the aged.* Monterey, CA: Brooks-Cole.
Easterlin, R. A. (1980). *Birth and fortune.* New York: Basic Books.
Elder, G. H., Jr. (1978). Approaches to social change and the family. In J. Demos & S. S. Boocock (Eds.), *Turning points* (pp. 1–38). Chicago: University of Chicago Press.
Elder, G. H., Jr. (1984). Family and kinship in sociological perspective. In R. Parke (Ed.), *The family* (pp. 426–451). Chicago: University of Chicago Press.
Erickson, E. H. (1950). *Childhood and society.* New York: Norton.
Esler, A. (1972). Youth in revolt: The French generation of 1830. In R. J. Bezucha (Ed.), *Modern European history* (pp. 141–161). Lexington, MA: D. C. Heath.
Featherman, D. (1982). Social stratification and mobility: Two decades of cumulative social science. *American Sociological Review, 24,* 364–385.
Featherman, D., & Lerner, R. M. (1985). Ontogenesis and sociogenesis: Problematics for theory and research about development and socialization across the lifespan. *American Sociological Review, 51,* 659–676.
Fendrich, J. M. (1974). Activists ten years later: A test of generational unit continuity. *Journal of Social Issues, 31,* 95–118.

Feuer, L. (1969). *The conflict of generations*. New York: Basic Books.

Flacks, R. (1967). The liberated generation: An exploration of the roots of student protest. *Journal of Social Issues, 23*, 52–72.

Freeman, J. (1975). *The politics of women's liberation*. New York: David McKay.

Friedan, B. (1963). *The feminine mystique*. New York: Dell.

Friedan, B. (1984). *Women of a certain age*. New York: Random House.

Friedenberg, E. (1969). Current patterns of generational conflict. *Journal of Social Issues, 25*, 21–38.

Gallagher, B. J. (1976). Ascribed and self-reported attitude differences between generations. *Pacific Sociological Review, 19*, 317–332.

Glass, J., Bengtson, V. L., & Dunham, C. (1986). Attitude similarity in three generational families: Socialization, status inheritance, or reciprocal influence. *American Sociological Review, 51*, 685–698.

Hagestad, G. O. (1977). *Role change in adulthood: The transition to the empty nest*. Unpublished manuscript, University of Chicago, Committee on Human Development.

Hagestad, G. O. (1981). Problems and promises in the social psychology of intergenerational relations. In R. W. Fogel, E. Hatfield, S. B. Kiesler, & E. Shanas (Eds.), *Aging: Stability and change in the family* (pp. 11–47). New York: Academic Press.

Hagestad, G. O. (1984). The continuous bond: A dynamic multigenerational perspective on parent-child relations between adults. In M. Perlmutter (Ed.), *Minnesota Symposia on Child Psychology* (Vol. 17, pp. 129–158). Princeton, NJ: Erlbaum.

Haraven, T. K. (1977). Family time and historical time. *Daedalus, 106*, 57–70.

Hernes, G. (1976). Structural change in social pressure. *American Journal of Sociology, 82*, 513–547.

Jennings, M. K., & Niemi, R. C. (1981). *Generations and politics*. Princeton, NJ: Princeton University Press.

Keniston, K. (1968). *Young radicals: Notes on committed youth*. New York: Harcourt Brace Jovanovich.

Kuhn, T. (1962). *The structure of scientific revolutions*. Chicago: University of Chicago Press.

Lerner, R. M. (1975). Showdown at the generation gap: Attitudes of adolescents and their parents toward contemporary issues. In H. D. Thornberg (Ed.), *Contemporary adolescence: Readings* (pp. 89–114). Belmont, CA: Wadsworth.

Lerner, R. M., & Knapp, J. R. (1975). Actual and perceived intergenerational attitudes of late adolescents and their parents. *Journal of Youth and Adolescence, 4*, 17–36.

Leventman, S. (Ed.). (1982). *Counterculture and social transformation: Essays in negativistic themes in sociological theory*. Springfield, IL: Charles C. Thomas.

Mannheim, K. (1952). The problems of generations. In D. Kecskemeti (Ed.), *Essays on the sociology of knowledge* (pp. 276–322). London: Routledge & Kegan Paul.

Mead, M. (1970). *Culture and commitment: A study of the generation gap.* New York: Longmans.

Miller, B., & Gilmore, R. (1975). *Revolt at Berkeley.* New York: American Publishing.

Nassi, A. J. (1981). Survivors of the sixties: Comparative psychosocial and political development of former Berkeley student activists. *American Psychologist, 36,* 753–761.

Nassi, A. J., & Abramowitz, S. I. (1976). Transition or transformation? Personal and political development of former Berkeley free speech movement activists. *Journal of Youth and Adolescence, 8*(3), 21–35.

Reich, C. (1970). *The greening of America.* New York: Dell.

Riley, M. W. (1985). Age strata in social systems. In R. Binstock & E. Shanas (Eds.), *Handbook of aging and the social sciences* (2nd ed., pp. 369–414). New York: Van Nostrand Reinhold.

Rossi, A. S. (1980). Aging and parenthood in the middle years. In P. Baltes & O. G. Brim (Eds.), *Life-span development and behavior* (Vol. 2, pp. 283–297). New York: Academic Press.

Rossi, A. S. (1987). Parenthood in transition: From lineage to child to self-orientation. In J. Lancaster, J. Altmann, A. Rossi, & L. Sherrod (Eds.), *Parenting across the life span* (pp. 31–84). Hawthorne, NY: Aldine De Gruyter.

Ryder, N. B. (1965). The cohort as a concept in the study of social change. *American Sociological Review, 30,* 834–861.

Sherif, M., Sherif, C., & Nebergall, R. E. (1965). *Attitude change: The social judgment-involvement approach.* Philiadelphia: W. B. Saunders.

Smith, T. (1983). Parental influence: A review of the evidence of influence and theoretical model of the parental influence process. In A. Kerkhoff (Ed.), *Research in sociology of educational socialization* (Vol. 4, pp. 13–45). Greenwich, CT: JAI Press.

Surrey, D. S. (1982). *Choice of conscience: Vietnam era military and draft resisters in Canada.* New York: Praeger.

Thompson, L., Clark, K., & Gunn, W. (1985). Developmental stage and perceptions of intergenerational continuity. *Journal of Marriage and the Family, 47,* 913–920.

Thompson, L., & Walker, A. J. (1984). Mothers and daughters: Aid patterns and attachment. *Journal of Marriage and the Family, 46,* 313–322.

Treas, J., & Bengtson, V. L. (1986). Family in later years. In M. Sussman & S. Steinmetz (Eds.), *Handbook on marriage and the family* (pp. 625–648). New York: Plenum.

Troll, L, & Bengtson, V. L. (1979). Generations in the family. In W. Burr, R. Hill, & I. Nye (Eds.), *Theories about the family* (Vol. 1, pp. 127–161). New York: Free Press.

Tuchman, B. (1984). *The march of folly: From Troy to Vietnam.* New York: Harper & Row.

Uhlenberg, P. (1978). Changing configurations of the life-course. In T. Haraven (Ed.), *Transitions* (pp. 66–98). New York: Academic Press.

Vinovskis, M. A. (1978). Recent trends in American historical demography: Some methodological and conceptual considerations. *Annual Review of Sociology, 4,* 603–627.

Whalen, J., & Flacks, R. (1982). The Isla Vista "bank burners" ten years later: Notes on the fate of student activists. *Sociological Focus, 13,* 215–236.

Wheeler, J. (1984). *The Vietnam generation*. New York: Random House.

Wood, J. L., & Ng, W. C. (1980). Socialization and student activism: Examination of a relationship. In L. Kriesberg (Ed.), *Research in social movements, conflicts, and change* (pp. 21–43). Greenwich, CT: JAI Press.

Yinger, J. M. (1982). *Countercultures*. New York: Free Press.

4

Productive Aging

Robert N. Butler

As former Senator Jacob K. Javits remarked in May 1984, two years before his death: "The most positive therapy is to perpetuate the life force, and whether the patient is a mechanic or U.S. Senator, he or she has a motivation which must prevail over the illness."

A somewhat different viewpoint was expressed by President Ronald Reagan who, also in 1984, told his 56-year-old challenger, Walter Mondale: "I want you to know that I also will not make age an issue in this campaign. I am not going to exploit for political purposes my opponent's youth and inexperience."

And to quote the subject of this festschrift, James Birren, at the Seventh International Congress of Gerontology in Vienna in 1966:

> With maturation comes a great conceptual grasp so that we can size up the situation and then look at the relevant items in our store. This is what I call the race between the chunks and the bits. While younger people, say those between 18 and 22, can process more bits per second, the older person may process bigger chunks. The race may go to the tortoise because

> he is chunking, and not to the hare because he is just bitting along. (Birren, 1966)

I first met James Birren in 1955. He was behaving like the younger person he was, processing many bits per second; but he was also an unusually mature person, ahead of his years, and processing big bytes. He understood early the importance of aging, and he understood it in the broadest sense. A true interdisciplinarian with the highest standards, he managed always to remain nondoctrinaire and open to ideas.

My purpose here is to explore productive aging, a subject that reflects so much of James Birren the man and his work. The beginnings can be found in his earliest work, on which we collaborated. It was James Birren who helped us to understand that disease had to be separated from aging and to realize the extraordinary potential that resides within the elderly. Part of the legacy of James Birren is the revision of the many myths and stereotypes related to aging that were common at the time of his, and perhaps our, earlier intellectual development.

One myth is that of aging itself, that is, chronological aging (the measure of one's age by the number of years one has lived). It is a myth because there are great differences in the rates of physiological, chronological, psychological, and social aging within an individual and from individual to individual (Butler, 1975).

Another myth is that of unproductivity. In fact, in the absence of diseases and social adversities, older people can and do remain productive and actively involved in life. The ability to change and adapt has little to do with one's age and more to do with one's lifelong character, so the common notion that increasing inflexibility comes with increasing age is another myth. There is, too, the myth of senility, that is, the idea that old people are senile, showing confusional episodes, forgetfulness, and reduced attention. In truth, much of what is called senile is a result of brain damage. Anxiety and depression are frequently put in the category of senility, even though they are treatable and often reversible conditions.

Finally, there is the myth of sexlessness, the idea that older people do not have the desire for or capability of a successful sexual life. These myths directly and indirectly affect the public's perception of older people's productivity and negatively influence older people's images of themselves and the contributions they can make to society.

After the first five years of our studies of human aging our broad conclusion could be stated as follows:

> As a consequence of a careful multidisciplinary pilot study, we have found evidence to suggest that many manifestations heretofore associated with

> aging *per se* reflect instead medical illness, personality variables, and sociocultural effects. It is hoped that future research may further disentangle the contributions of disease, social losses, and preexistent personality, so that we may know more clearly what changes should be regarded as age-specific. Indeed, various types of investigations complementing one another would be useful. Intensive studies, involving frequent contact over considerable periods of time, based upon the growing personal relationship between the investigator and the older person, would contribute to our understanding of the subjective experience of aging and approaching death. . . . Longitudinal studies, of course, would enhance our opportunities of classifying changes as to whether they are age-specific, disease-linked, etc. If we can get behind the facade of chronological aging, we open up the possibility of modification through prevention and treatment. In our lifetime (if at all) it is not likely that the inexorable processes of aging will be amenable to human intervention, but it cannot be too greatly emphasized that it is necessary to be able to recognize those factors which are open to change. (Butler, 1963, p. 728)

I cannot speak for Birren, but I do not recall having at the time a clear understanding of the link between our research findings and their application to public policy. I remember feeling that it was inaccurate and discriminatory to regard older persons as unproductive; in addition I saw the issue more in terms of the individual than as one of needed public policies to assure society of the productive contribution of older persons.

Our multidisciplinary team took 8 years to complete our book (Birren, Butler, Greenhouse, Sokoloff, & Yarrow, 1963), the time factor probably being a reflection of the pleasure we took working together. This period was truly one of the highlights of my intellectual life. I cannot help but recall the wonderful time we had in Chicago in 1967, when there was a great blizzard and, happily, we were "stuck together" for a 4-day weekend and discussed what we wished to see happen in the field of gerontology in the ensuing years. Among some other things, Birren and I requested a laboratory of aging at the National Institute of Mental Health in 1963; we wanted to create a major laboratory and address the problems of depression and dementia. Unfortunately we were turned down. I have always regretted this outcome because I believe we would have worked remarkably well together as co-directors of the laboratory.

But enough of reminiscences. I will treat my topic more in the form of an essay than a formal review, but I will refer prinicipally to the scientific basis for productive aging and its public policy aspects. The assumption is that to be productive, to have substantial purpose, a sense of meaning is essential to the human experience. Productive aging is the critical response to the worldwide longevity revolution. Extended life expectancy necessitates an extended work life, paid and unpaid.

Productive aging is a relatively new issue that represents a potential solution to several other issues that I will describe. I think it is a concept that developed from a political and socioeconomic basis and is rather attractive to people all along the political spectrum. Some of the thinking that I am going to describe—some of the chunks and bits—derives from experiences Birren and I had together at the Salzburg Seminar a few years ago (Butler & Gleason, 1985) and from work at the Center for Productive Aging at Mount Sinai.

There are three great concerns, even fears, about old age that I have encountered in every society I have had occasion to visit—both as an official guest during my tenure as director of the National Institute on Aging and as a private citizen—regardless of the political system and the socioeconomic arrangement of that society (e.g., the Soviet Union, People's Republic of China, Japan, Romania, and France). With the remarkable increase in absolute numbers and proportions of older persons come these three great concerns: (1) the burden and cost of old age, (2) the prospect of intergenerational conflicts, and (3) the notion of stagnation or senility of societies. The last is an old French worry, most recently expounded by Alfred Sauvy (1969) but actually going back to the last century, when France began to realize that it was the first of our aging societies. All three of these fears or concerns are interconnected. What can we do? There are three responses that bear upon the above three concerns: science, policy, and attitudes.

THE COST OF OLD AGE

First let us take up the issue of the enormous cost of Social Security and health care for older people. We clearly have to implement some major cost and quality reforms in our health care system. To accomplish this, we must take a very different approach to medicine than we presently do, one that creates a very different social organization. This means the end of the physician as the elitist who authoritatively controls the patient and his or her care. Instead we must organize a team that builds an egalitarian relationship among nurses, social workers, physical therapists, clinical pharmacists, speech pathologists, and physicians. This notion of the interdisciplinary or multidisciplinary team grows out of my own sensitization in working with Birren on our project and in a multidisciplinary lab. There too, we were very much interested in the emphasis upon function. It is not enough to be able to do something about a specific disease if the patient's functional status is not improved.

We must pay attention to the great antecedents to institutionalization: first, problems of mobility; second, memory; third, incontinence. *Problems of mobility*—gait disorders, and falls—constitute the number-one category of causes for admission to nursing homes and other institutions. *Problems of mobility* also often bring an end to independence. Interdepartmental programs to effectively diagnose, prevent, and treat such problems are badly needed. For example, neurology, rehabilitation, and geriatrics services could combine and organize effective clinics that apply new knowledge from studies of falling and gait.

Although we are a long way from cure, *problems of memory* can be ameliorated. There is an increasing need for programs for the early diagnosis of Alzheimer's disease and related disorders. *Urinary incontinence* is especially common in women who have borne children, after they enter their fifties, and it also occurs in men. This is a socially disabling problem, but there are important therapeutic approaches—mechanical, pharmacological, and surgical—which can be effective. These three problems effectively illustrate the importance of function that is so central to high-quality geriatric practice.

Furthermore, we need to restructure Medicare, which was built originally as if people 65 and over were really only 40, that is, without attention to physical checkups, outpatient medication, long-term care, home care, prosthetics, and rehabilitation. Medicare was a financing approach developed in the absence of a functioning relationship between the private and public sector. And while Medicare contributes enormously to graduate medical education in the United States, it has never contributed a nickel to the development of geriatrics. A restructuring is necessary to help deal with the tremendous cost burden.

Now for a scientific strategy. Birren took part in our effort at the National Institute on Aging to create a research plan—*Toward an Independent Old Age—A National Plan for Research on Aging* (1982)—that now needs updating, prioritizing, and pricing out. The roles of aging in human affairs and of aging as one of the three antecedents of disease have to be studied more intensely than they presently are. Along with genetics and environment, broadly defined from pathogens to life-style, aging is an antecedent to disease but has been the stepchild in terms of its role in the genesis of disease. We must support longitudinal studies that are truly representative and comprehensive, exploring health and creativity and not just disease. We have spent extraordinarily little money in such research to date in this and most other countries. We need a national populations laboratory.

In the follow-up to the human aging work after the eleven-year study, where none of the different disciplines was aware of the findings of the other, we found to our surprise and interest that survival was very much heightened in those individuals who had specific goals and a sense of order and organization in their everyday life. As we move toward the next century and see the extraordinary change in the absolute number and relative proportion of older persons, we face a choice between having an expanding dependent class and maintaining productivity. To choose the latter we will need to know more about human performance in a variety of conditions. This means we must study people throughout life, well into their sixties, seventies, and eighties. We have developed an understanding of the importance of following the natural course of diseases, but we know very little about the natural course of lives. We require a much better understanding than we have today of the relationship between work ability and chronological age. As a result of the veritable explosion of understanding in the neurosciences, as well as the modern technology available to us, and with the realization of the remarkable possibilities of continuing repair and growth in central nervous system functioning, we have the chance to gain a much clearer perception of the continuing nature and change of human performance.

If Alzheimer's disease goes unchecked because we do not discover its character and genesis, we will face a serious problem. As Jacob Brody (1985) predicted, by 2050 we could have a staggering 9 million cases of Alzheimer's disease. Fries (1980) predicted a reduction in morbidity and health costs and a slowing down in the rate of increase in the over-85 population. However wrong his predictions, the ultimate goal of maintaining vigor and reducing morbidity for as long as possible is certainly the appropriate strategy. To that end, the ultimate cost containment and the ultimate service is research itself.

And what about attitudes? I grew up in a medical environment and was horrified by such epithets as "crock" applied insensitively to middle-aged women and older people. I am reminded of the novel *The House of God* by Sam Shen (1978), about a young physician who details the use of this unfortunate term used sotto voce against older people, reflecting, I think, the sense of helplessness or impotence we feel in trying to help patients with complex and multiple problems. We must also begin to think in terms of the vigorous and continuing health promotion and disease prevention which provide each of us with further control over our own lives. Aging is a highly individual process, something that Birren has taught us.

INTERGENERATIONAL CONFLICT

Now for that second concern, intergenerational conflict, which has recently been a preoccupation of the media. It is a little like instigating gang wars and a kind of remake of the "generation gap" of the 1960s. The emotionally explosive rhetoric includes phrases such as "mortgaging the future." The old and the young are not natural enemies. I submit that the enemies are misinformation, ideology, and competing interests of another sort. And there is much confusion in the presentation of the finest of our media, suggesting, for example, that the dependency ratio is rising and will continue to do so in the next century. In reality, there is a continuing drop in the dependency ratio, and a policy analysis of the cost of raising a child compared with the cost of caring for an older person still remains to be done.

Moreover, most discussions of income transfers have centered on Social Security, suggesting that most transfers go from young to old. But some studies from the University of Michigan looking at all income transfers from the private as well as the public sector show, rather, that income and assets move from old to young. Social Security and pension fund reformation become very important. From the beginning, Social Security was not a pension plan for older people; rather it was supposed to be a life-cycle contingency fund concerned with disability and survivorship as well as retirement. It should, however, be built as a life-cycle and intergenerational program. This would help to offset the tendency to generate strife among the generations. We also must note the issue of private pensions—extraordinarily huge amounts of money that remain essentially underregulated and often do not provide early and adequate vesting. The Tax Reform Act of 1986 moves vesting down to 5 years from 10 and offsets the present technique of integration of Social Security with private pensions, which will greatly augment the economic base and retirement of American workers.

The notion of national health insurance, presumably dead, is coming back. The issue of long-term care in particular has begun to stimulate renewed interest in the participation of the public sector in the provision of national health insurance. The annual average cost for nursing home care is $22,000 per year, out of reach of the great majority of people. This can be a major catastrophe for families. In order to gain access to financial coverage, one has to go through the humiliation of "spending down" of assets and income to gain eligibility under Medicaid. Physicians who had strongly opposed national health insurance are beginning to feel that they would rather negotiate with government than work for corporations. And former Health and Human Services

Secretary Joseph Califano estimated some $2 trillion as the unfunded retiree health care liability in the United States, a sum that makes it extremely difficult for this country to compete internationally. So it may now be in the interest of medicine, business, and the American citizen, in terms of asset protection, to move toward a true national health insurance. And we may need to specify an international minimum wage because labor costs are determined by multinational corporations that move part of their work to countries whose workers are underpaid, destroying the carefully designed welfare systems of advanced nations.

STAGNATION OF SOCIETY

The third concern is the stagnation of society. We must look at productivity as a whole and the role of the aged within it. We need only go through a city and see the physical decay, the slums, and know that what is needed is an overarching effort to match skills and needs. We may have to reinvent the CCC (the Civilian Conservation Corps) for the young and the old and reinstitute it at community, state, and federal levels—an effort that would surely help to mobilize and continue the productive contributions of older and younger citizens. I have also repeatedly proposed the development of a National Senior Service Corps which would provide an opportunity for continuing contributions, or a National Service Corps with a blend of all ages. Older persons have made distinguished contributions to VISTA, the Peace Corps, Green Thumb, the Foster Grandparents program, and similar public as well as private sector organizations. There is a natural nexus between health and productivity. And instead of further confining life's key activities to three separte areas—education, work, and retirement—we need to interweave them in ways that are constructive and creative as regards our society as well as the individual, resulting in a continuing investment in human development, for children as well as older persons.

Sweden has a Ministry of the Future. There is no such cabinet department in our country. Our productivity, our use of people, is organized around "the bottom line," no longer by corporate annual reports but now even by quarterly earnings reports. We will have to transcend the limitations of our thinking and our national bookkeeping, our accounting. When we talk about budget deficits, we should rethink our concept of the budget. We need a *productivity budget* which includes the development of human beings, their education and good health. We need a productivity budget that exploits the marvels of biotechnology and high technology. We should follow the example of Japan and the

People's Republic of China and set up "science cities," where a high concentration of technological talent creates jobs for those who are economically less fortunate. Productivity, then, has to be seen as a whole. Let us no longer be concerned with such issues as, will older people who stay in the work force take away jobs from the young? The work force is not that clearly interchangeable. The International Labor Office in Geneva, in fact, has, despite its numerous studies of various societies, been unable to show that the presence of older people in the work force has taken jobs away from the young.

CONCLUSION

This is the century of old age, the first century in which older people are mass-produced. What has been the privilege of the few has become the destiny of the many. The advanced and developing nations have experienced remarkably increasing absolute numbers and relative proportions of older persons. This is the longevity revolution. We have gained 25 years of life expectancy in less than a century—nearly the same amount that we gained in the previous 5,000 years. We have come a long way from that brutish short life that Hobbes described in *Leviathan*, thanks to the decreased maternal, childhood, and infant mortality rate, and, more recently and happily, to the marked reductions in deaths from heart disease and stroke. Death is now the business of old age: 80% of all deaths occur after age 60. According to demographic projections, by the year 2000 there will be 100,000 centenarians. But listen to this: in less than 50 years, by the year 2028, with only a 2% drop in mortality rates per year, we could have 19 million centenarians. Unless, of course, we slow down science and the work at the National Institutes of Health. The past and present show us the great possibilities of survival, and at the same time present us with a challenge, a social and public health challenge, that requires new mindsets: a transformation of culture, society, and socioeconomic arrangements; a transformation in education, work, and retirement arrangements; a movement toward productive aging.

Productive aging includes continuing growth and development and affords us the means to allay some of our concerns relating to the triple threat of old age: runaway costs, intergenerational strife, and stagnation. Life is really of one piece, a whole series of marvelous transformations throughout the cycle, a great gift, and a remarkable opportunity for joy and warmth and human creativity.

REFERENCES

Birren, J. E. (1966). The regulation of aging and behavior by the nervous system. *Proceedings of the Seventh International Congress of Gerontology*, 299–300. Vienna: Verlag der Wiener Medizinischen Akademie.

Birren, J. E., Butler, R. N., Greenhouse, S. W., Sokoloff, L., & Yarrow, M. R. (Eds.). (1963). *Human aging I: A biological and behavioral study*. Washington, DC: U.S. Government Printing Office. (Public Health Service Publication No. 986; reprinted in 1971 and 1974)

Brody, J. (1985). Prospects for an aging population. *Nature, 315*, 463–466.

Butler, R. N. (1963). The facade of chronological age: An interpretative summary of the multidisciplinary studies of the aged conducted at the National Institutes of Mental Health. *American Journal of Psychiatry, 119*, 721–728.

Butler, R. N. (1975). *Why survive? Being old in America*. New York: Harper Torchbooks.

Butler, R. N., & Gleason, H. P. (1985). *Productive aging, enhancing vitality in later life*. New York: Springer Publishing Co.

Fries, J. F. (1980). Aging, natural death and the compression of morbidity. *New England Journal of Medicine, 303*, 130–135.

Sauvy, A. (1969). *General theory of population*. New York: Basic Books.

Shen, S. (1978). *The house of God*. New York: Dell.

Toward an independent old age—a national plan for research on aging. (1982). Washington, DC: National Institute on Aging.

5

Individual Differences in Rate of Cognitive Change in Adulthood

K. Warner Schaie

Throughout his illustrious research career, James Birren has been in the forefront of those who have insisted that the study of group trends in aging phenomena must be accompanied by careful analyses of individual differences in age trends (Birren, 1959). He pioneered work that dealt with the fact that aging effects might differ markedly in healthy populations or those at risk (e.g., Birren, Butler, Greenhouse, Sokoloff, & Yarrow, 1963). A number of empirical studies conducted by Birren, moreover, dealt directly with individual difference variables in the study of human intelligence (e.g., Birren & Morrison, 1961). It seemed most appropriate, therefore, that my contribution to this volume should examine in some detail the role individual differences come to play in the process of cognitive change across the adult life span.

The research reported in this chapter was supported by a research grant (AG04770) from the National Institute of Aging. The continuing support of members and staff of the Group Health Cooperative of Puget Sound, our sampling frame, is gratefully acknowledged. Thanks are also due to Cherill Perera, who supervised the data collection, and Ann O'Hanlon, who supervised the data reduction.

Extensive research on psychometric intelligence over the adult life course has made us familiar with the finding that most abilities tend to peak in early midlife, plateau until the late fifties or early sixties, and then show decline, initially at a slow pace, but accelerating as the late seventies are reached. Some controversy remains on the specific ages at which certain abilities peak and on the ages at which significant decline can first be detected (cf. Botwinick, 1977; Labouvie-Vief, 1985; Willis, 1985).

Data from cross-sectional studies typically result in more pessimistic findings for variables for which positive cohort trends have been observed (that is, where later-born individuals perform at a higher level at the same age than do earlier-born individuals) and unduly optimistic findings for those variables where cohort trends have been negative (that is, where earlier-born individuals performed at a higher level at the same age than do later-born individuals). Most age-comparative work is flawed also because of the fact that it is difficult or impossible to match samples differing widely in age with respect to other variables that might critically affect the dependent variable of interest. Longitudinal studies control for cohort effects and of course provide within-subject comparisons. Nevertheless, they may provide under- or overestimates of mean age changes, depending upon whether they are favorably or unfavorably attrited (Schaie, 1983a; Schaie & Hertzog, 1982).

While the above-mentioned research literature has been useful in informing us on group trends, such trends do not necessarily represent the patterns of cognitive change for all or most individuals. Some of the other chapters in this volume center about models of biological, social, and psychological aging that would help us understand the aging process in broad strokes. Such models are certainly useful, but the vast individual differences observed in all such processes require us in addition to consider the many variations in which individual aging may be expressed (cf. Birren & Cunningham, 1985; Birren & Renner, 1977). Although a cumulation of deleterious age changes is inevitable for all of us as we age, there are many factors, whether genetically or environmentally programmed, that influence the rate at which age changes occur. For a full understanding of these issues it is necessary to engage in a process of successive disaggregation that will lead us to understand how individual aging may differ from the group norms that we often must rely on for purposes of policy formation.

The specific purpose of this chapter, therefore, is to take a step forward in advancing the study of cognitive development in adulthood beyond the analysis of average group profiles that describe differences in behavior between or within groups of individuals differing in age.

For this purpose, some detailed analyses will be presented on those participants in the Seattle Longitudinal Study (SLS) for whom data are now available for five measurement points over a period of 28 years. These longitudinal data not only allow comparison of levels of performance within individuals but also permit fitting slope parameters that will inform us on rates of change for individuals and groups having common characteristics. Three major cognitive abilities will be considered: Verbal Meaning, Spatial Orientation, and Inductive Reasoning. These particular variables were chosen because they are essential skills in communicating with others, in moving about the environment, and in virtually all problem solving involved in daily living (Willis & Schaie, 1986).

To be presented first are traditional data on group means to show that reliable average age changes indeed do occur in this data set. Next these data will be disaggregated into three subsets to show that average age changes over the past 28 years differ when considering those who are now middle-aged, young-old, or old-old. Average rates of change, obtained by fitting the linear slopes for each of our study participants, will then be examined. Types of change patterns will be examined that were obtained by clustering study members, and finally some individual profiles will be presented that call attention to the fact that there are wide individual differences in both slope and direction of age changes in cognitive behavior.

CHARACTERISTICS OF THE DATA BASE

The Subject Population

Some 30 years ago I began my inquiry into adult cognitive functioning by drawing a sample of 500 subjects evenly distributed by sex and age across the range from 20 to 70 years, by means of randomly sampling from the approximately 18,000 members of a health maintenance organization in the Pacific Northwest. The sample represented a broad distribution of educational and occupational levels, probably well representing the upper 75% of the socioeconomic spectrum. Of the original sample, 88 persons participated at all assessment points. The residual sample consists of 40 men and 48 women with an average age of 68.8 years (SD=10.3; range 50 to 95). All participants were community-dwelling and were not suffering from any acute disease as reported by their health plan physician.

The Measurement Variables

The variables to be discussed here involve the first three Primary Mental Abilities identified by the Thurstones (1941). They were assessed with the SRA Primary Mental Abilities Test (Schaie, 1985; Thurstone & Thurstone, 1949). *Verbal Meaning* involves the ability to recognize and comprehend words; it is a measure of a person's recognition or passive vocabulary. This ability is assessed by providing individuals with a stimulus word and asking them to match it from a multiple choice list. Verbal Meaning is thought to be a crystallized ability that is acquired and maintained by exposure to culture-determined experiences. *Spatial Orientation* is the ability to rotate objects mentally in two-dimensional space. This ability is involved, for example, in visualizing the direction one might enter a highway after having inspected a map, or in assembling a piece of furniture by following a set of instructions. Spatial Orientation is measured by providing the subject with an abstract stimulus figure and then asking the subject to select rotated examples of that figure that would match the stimulus upon mental rotation. *Inductive Reasoning* is the ability to identify regularities and to infer principles or rules. It is a critical component of most problem-solving tasks. This ability is measured by asking subjects to complete a letter series task. Both Spatial Orientation and Inductive Reasoning are thought to be fluid abilities that are involved in the mastering of novel experiences. All of the measures are slightly speeded paper-and-pencil tasks.

The Assessment Procedure

The measures just described were administered to small groups of subjects as part of a broader test battery that originally required 2 hours in a single session but has since grown to a 5-hour battery spread over two sessions. The Primary Mental Ability measures, however, have always been given in the same position during the first part of the testing session. Subjects were first tested in 1956 and were retested in 1963, 1970, 1977, and 1984–1985.

FINDINGS ON AVERAGE AGE CHANGES

We begin our exploration by examining the progression of our subjects across the 28 years of our study. As will be noted from Figure 5.1, the average ages of our residual sample at the five measurement points were 41, 48, 55, 62, and 69 years, respectively. For convenient comparison across the three ability measures, all raw scores were rescaled in T-score form (M = 50, SD = 10), using the parameters obtained for a broad sample of 2,810 subjects at first test with an average age of 53 years (Schaie, 1983b). Scaling in this fashion not only permits us to examine change within our long-term panel but also helps us to understand the magnitude of change by placing it within the context of a broader population frame.

As would be expected in a favorably attrited panel, our residual subjects on average were about one half of a standard deviation (*SD*) above the population mean when they were first studied; by the last measurement point they are still slightly above the population mean. For the total sample statistically significant decline can be detected by age 62. Verbal Meaning actually continues to increase slightly until age 55. The decline by age 69 amounts to .33 *SD* from the initial level, and .46 *SD* from peak level. Spatial Orientation appears to show decline by age 55; however, 7-year age changes on this variable are not statistically reliable until age 69 is reached, although cumulative decline reaches significance by age 62. The magnitude of change for Spatial Orientation

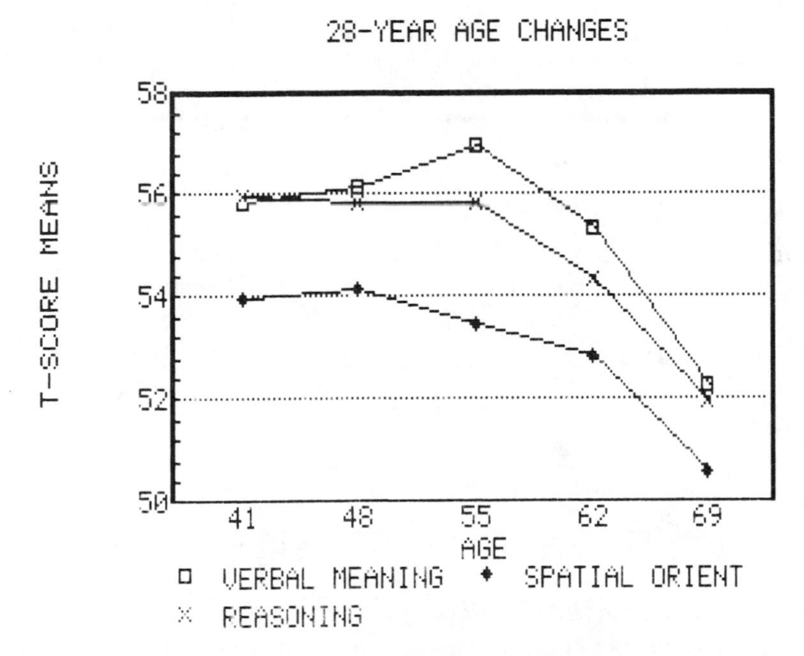

FIGURE 5.1 Age changes over 28 years for the total sample for the abilities of Verbal Meaning, Spatial Orientation, and Inductive Reasoning.

amounts to .34 *SD* from initial level. Inductive Reasoning remains level until age 55 and then declines, the magnitude of decline amounting to .40 *SD*. Fitting a straight line through the average age trends indicates that the annual average rate of decline over the age range from 41 to 69 years amounts to .012 *SD* for Verbal Meaning and Spatial Orientation and to .014 *SD* for Inductive Reasoning.

Age Changes Differentiated by Cohort Level

Because of the wide age/cohort range represented in our sample, these age trends may conceal important differences in change patterns for successive cohorts and different age ranges. The data were therefore disaggregated first into three subsamples, those who at the last time of measurement were middle-aged ($N = 30$; age $M = 57$, $SD = 3.5$), young-old ($N = 39$; age $M = 71$, $SD = 3.5$), and old-old ($N = 19$; age $M = 85$, $SD = 3.9$). This disaggregation yields longitudinal segments that cover the age range from 29 to 85, with two cohort comparisons at ages 43, 50, 64, and 71 and three cohorts compared at age 57. Figures 5.2, 5.3, and 5.4 present these data by ability.

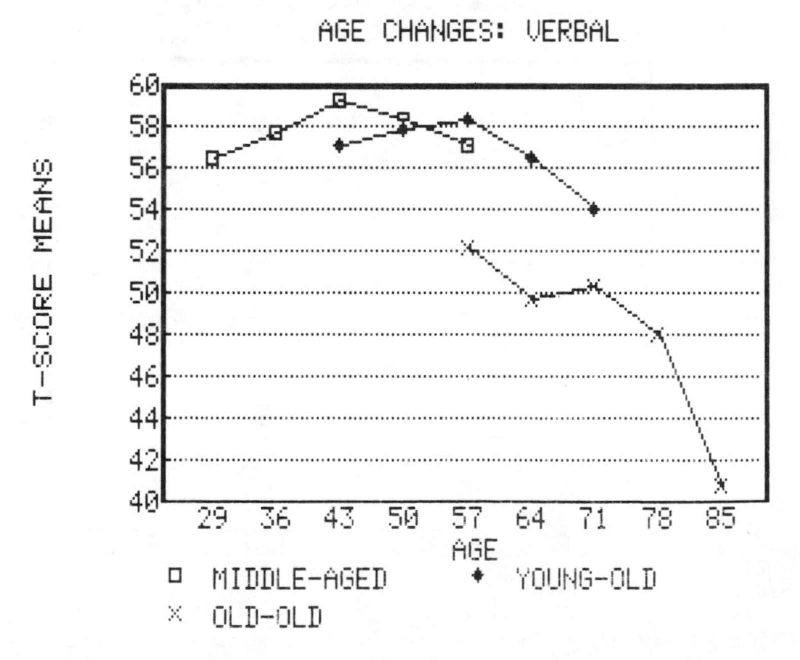

FIGURE 5.2 Age changes over 28 years for Verbal Meaning.

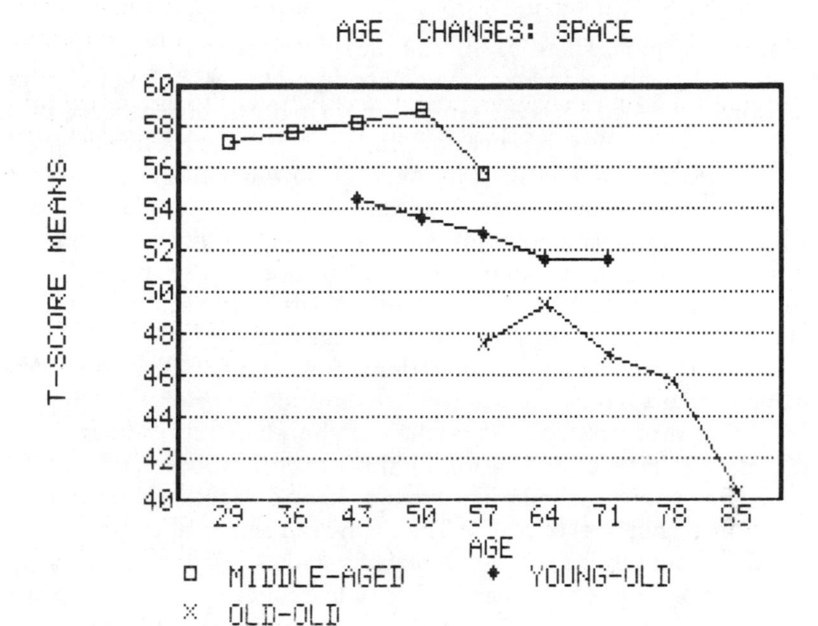

FIGURE 5.3 Age changes over 28 years for Spatial Orientation.

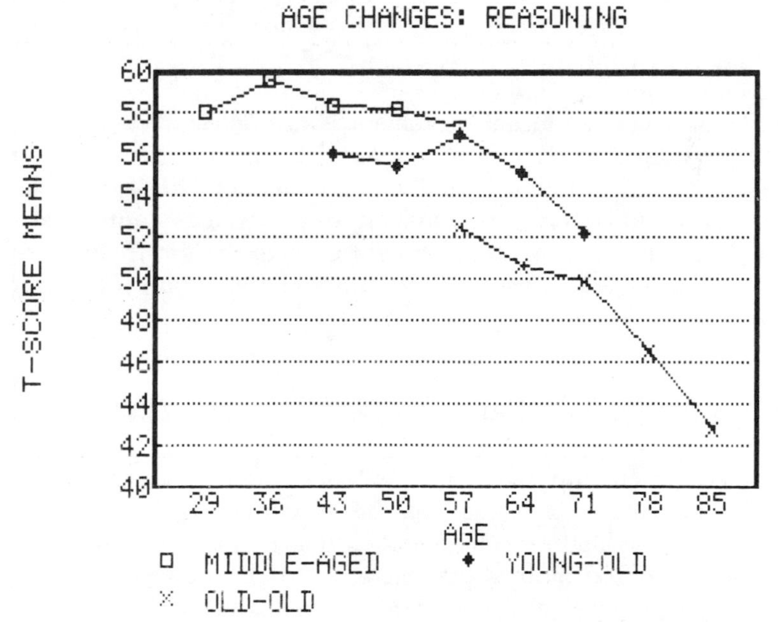

FIGURE 5.4 Age changes over 28 years for Inductive Reasoning.

The data for Verbal Meaning (Figure 5.2) reflect interesting age-by-cohort interactions. There is a substantial difference in favor of the middle cohort over the oldest cohort at all comparable ages. The peak age of performance (age 57) is the same for both cohorts, but the slope over these ages is slightly steeper for the middle cohort. Interestingly enough, the peak age for the youngest cohort actually occurs at age 43. For that cohort there is a statistically significant cumulative increment to age 43 and a cumulatively significant decrement from that point to age 57. The young-old cohort, by contrast, does not show significant decline until age 64. And for the oldest cohort, statistically significant cumulative decline is observed only by age 78.

The cohort pattern for Spatial Orientation (Figure 5.3) is more systematic in that each successive cohort is found to perform at a higher level at all comparable ages. Nevertheless, this ability also shows different peaks and rates of decline for the three cohort groups. The middle-aged cohort shows a shallow rise to age 50 and decline (although not statistically significant) by age 57. The young-old cohort appears to show slight decrement throughout, that reaches statistical significance by age 64 but then remains level to age 71. The oldest cohort, by contrast, peaks at age 64 and shows statistically significant decrement only by age 78.

The pattern of higher-performing successive cohorts also holds for Inductive Reasoning (Figure 5.4). The youngest cohort appears to peak at age 36; however, age changes for this group are not statistically significant. The young-old cohort peaks at 57, with statistically significant decrement first detected at age 64. The oldest cohort also peaks at 57, but statistically significant decrement is not detected until age 71.

Magnitudes of average decline are virtually trivial for the middle-aged cohort, are quite modest for the young-old cohort, but are substantial for the old-old cohort. These magnitudes as well as annualized rates of change obtained by fitting straight lines to the group averages are provided in Table 5.1 from the initial measurement point and from peak performance levels.

INDIVIDUAL DIFFERENCES IN RATE OF CHANGE

A much more complicated story unfolds when we begin to examine individual patterns of change over the 28-year period. For all three abilities and across all three cohorts individuals may be found whose slope coefficients reflect more severe decrement than the group average or, by contrast, whose performance over time proceeds in a positive fashion. For example, the annual linear rate of decline on Spatial Orienta-

TABLE 5.1 Magnitude of Age Changes in Standard Deviation Units by Ability and Cohort

	Verbal Meaning	Spatial Orientation	Inductive Reasoning
		From initial level	
Middle-aged	+.07	−.15	−.07
Young-old	−.30	−.29	−.39
Old-old	−.73	−.74	−.96
		From peak level	
Middle-aged	−.20	−.30	−.23
Young-old	−.41	−.29	−.48
Old-old	−.73	−.92	−.96
		Annual rate of change	
Middle-aged	+.002	−.003	−.005
Young-old	−.010	−.011	−.012
Old-old	−.035	−.026	−.033

tion for our subjects as a group was estimated to proceed at the rate of .012 *SD* units. The range of individual slope coefficients, however, ranges from a maximum annual decline of .101 *SD* to a maximum annual gain of .027 *SD* units. In other words, at least one of our subjects declined approximately eight times as fast as the group average, and others showed ability gains rather than decline.

Cohen (1977) has suggested that effect sizes expressed in population standard deviation units should be considered small (of little interest) if they are less than .2 *SD* units, of moderate importance if they reach .5 *SD* units, and of substantial interest if they exceed 1 *SD*. Employing these criteria we find that approximately 32% of our subjects show moderate or large decline on Verbal Meaning; similar figures are 34% for Spatial Orientation and 37% for Inductive Reasoning. To put it more positively, these data imply, of course, that approximately two-thirds of our participants showed little or no decline. Figure 5.5 employs the Cohen criteria in a more detailed breakdown by cohort for the ability of Spatial Orientation. Note that there is a virtually normal distribution of gain, stability, and loss for the middle-aged cohort, increasing loss but still marked stability for the young-old, but moderate to substantial loss for most of the old-old.

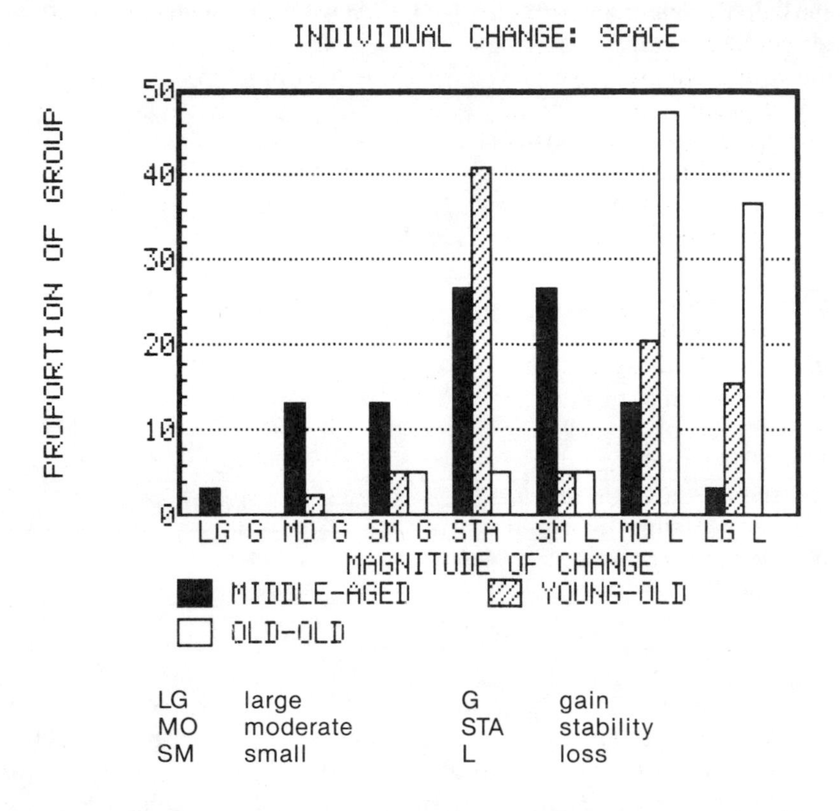

FIGURE 5.5 Proportion of subjects by cohort who gained, remained stable, or lost performance skills on Spatial Orientation.

Typical Patterns of Change

The above demonstration of wide individual differences in rate of change needs to be extended further to attend to the substantial individual differences in patterns of change. Some individuals, of course, remain stable throughout. Others show an early decline, followed by recovery to the prior level of functioning, whether by serendipitous events or as the intended consequence of programmed intervention (cf. Schaie & Willis, 1986). Some persons decline at a regular rate from young adulthood on; others show early decline, followed by a late life plateau, or they begin to show decline only quite late in life.

Our next step in the process of disaggregating the group data, therefore, consisted of conducting a cluster analysis (Spath, 1980; Ward,

1963) on the 39 members of our young-old cohort to determine whether we could develop a reasonable typology of profile types. We selected the young-old group because it appeared to contain the greatest diversity of profiles, including all of the alternatives discussed earlier. The clustering algorithm was first used to assign each of our subjects to one of two major profile types. Further clustering then resulted in assignment to one of four subtypes. Figure 5.6 gives an example of the tree-structure resulting from the cluster analysis for Spatial Orientation.

Major profile types. The clustering algorithm employing the Euclidian distances between all pairs of profiles yields an admixture of level and shape of these profiles. The following figures consequently compare the average profiles for the high-scoring and low-scoring members of the young-old cohort as compared with the profile for the entire cohort. The major profile types emphasize level differences but also call attention to the interaction between level of cognitive function and long-term change. This interaction is particularly noteworthy for Verbal Meaning (Figure 5.7). Note the group profile that peaks at age 57

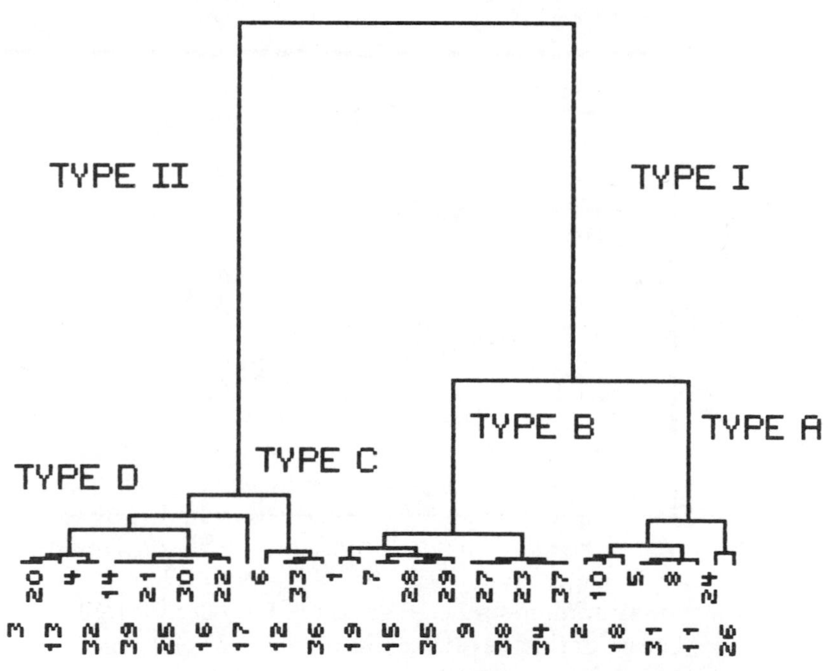

FIGURE 5.6 Tree diagram for the cluster analysis of Spatial Orientation data for the young-old cohort.

and declines thereafter. This group trend is replicated in more pronounced form by Type I, which includes most of the higher-performing individuals, with a virtually shallow profile until age 64 and only minimal decline thereafter. The average annual rate of change for Type I is almost three times as large as for Type II (.016 *SD* vs. .006 *SD*).

For Spatial Orientation (Figure 5.8), the group profile suggests steady decline until age 64 with a plateau over the next 7-year period. When disaggregated, the high-scoring individuals plateau until age 57, decline to age 64, then plateau again. The low-scoring individuals, on the other hand, show early decline by age 57 and a lower plateau thereafter.

For Inductive Reasoning (Figure 5.9), the group profile of modest increment to age 57 and decline thereafter is virtually replicated by the high-scoring members of the group. By contrast, low-scoring individuals appear to plateau in midlife and decline thereafter.

Rates of decline for Space and Reasoning do not differ as much by type as for Verbal Meaning. Nevertheless, in both instances the rate of the lower-scoring type is about 1½ times that of the higher-scoring type.

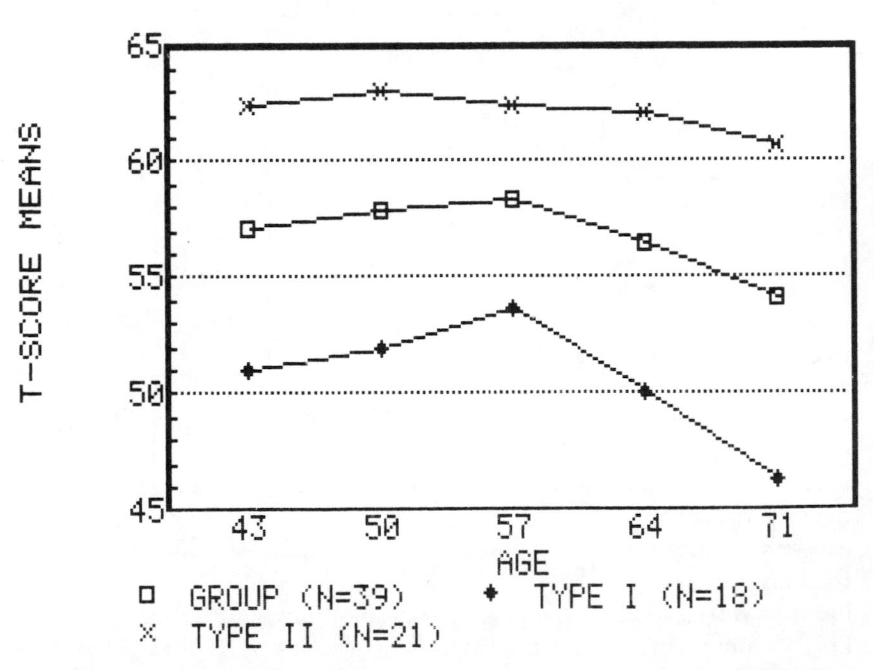

FIGURE 5.7 Mean profiles for group average and major types for the young-old cohort on Verbal Meaning.

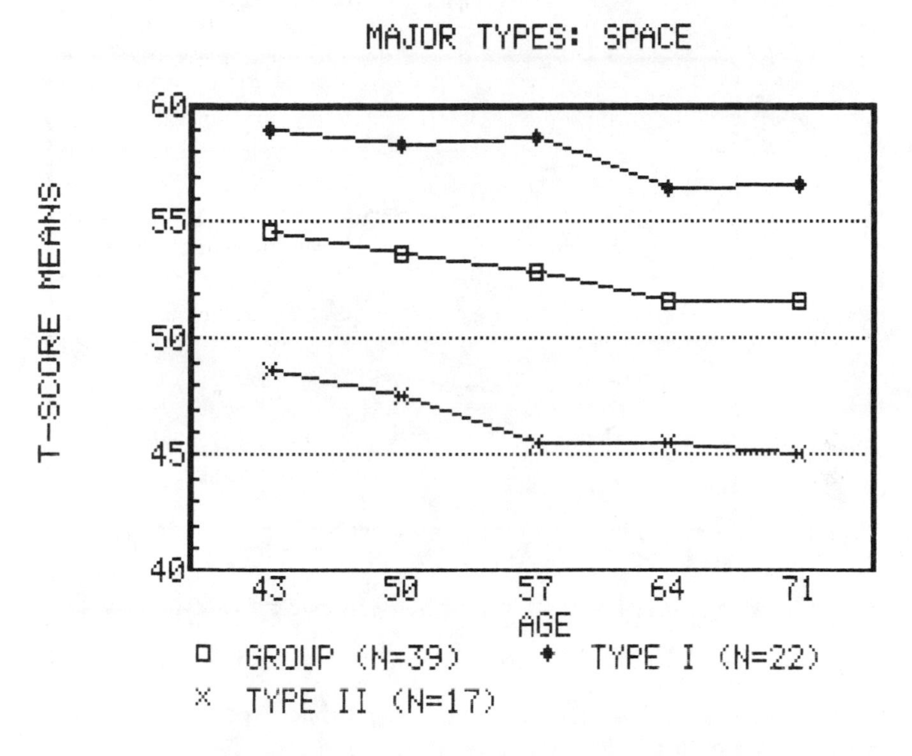

FIGURE 5.8 Mean profiles for group average and major types for the young-old cohort on Spatial Orientation.

Subtype profiles. As a next step we identified four subtype profiles for each of the three abilities. These subtypes come considerably closer in reproducing the variety of individual change patterns present in our data, while smoothing some of the irregularities in the individual data. For Verbal Meaning (Figure 5.10), the subtypes primarily identify different profiles among high-scoring individuals (Subtype A is virtually identical with major Type I). Subtype B represents a pattern of early decline, by age 56, and subsequent recovery by age 71. Subtype C replicates the group norm; i.e., peak attainment at age 57 and decrement thereafter. Subtype D, by contrast, represents virtual stability from midlife into early old age. The average rate of change for subtypes B and D is virtually zero, that for Subtype C is modest (.011 *SD*), while the rate of change for Subtype A is substantial (.022 *SD*).

Subtype profiles for Spatial Orientation (Figure 5.11) appear to be ordered

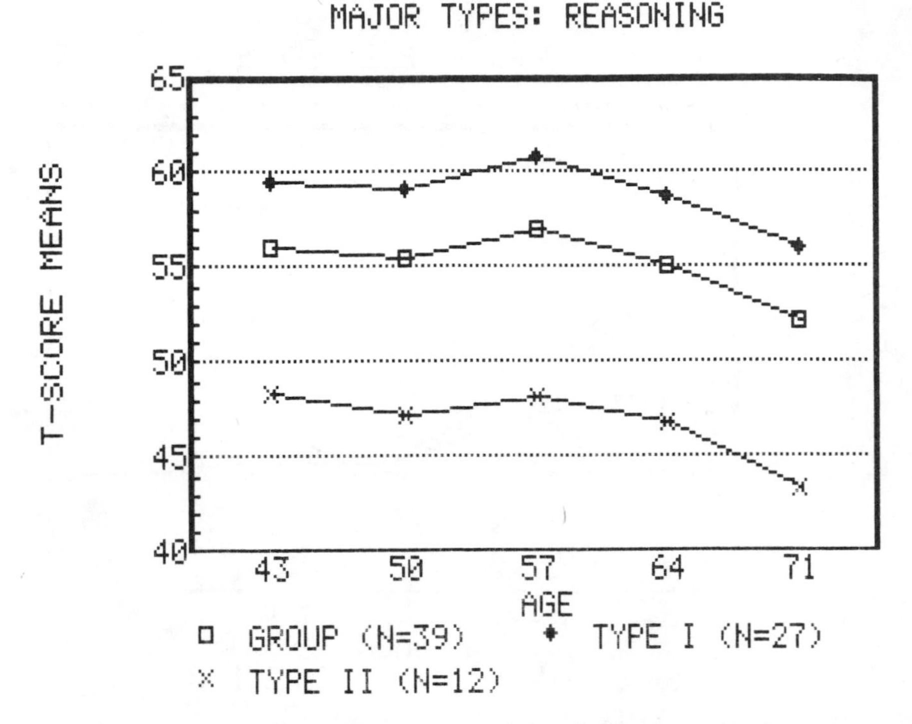

FIGURE 5.9 Mean profiles for group average and major types for the young-old cohort on Inductive Reasoning.

by performance level but also represent differences in change pattern. A linear fit for Subtype A suggests a positive trend throughout (annual change = +.008 *SD*), while Subtype B represents individuals declining at a steady rate from middle adulthood (annual change = .002 *SD*). Subtype C represents early decline, to age 57, followed by a plateau, while Subtype D reflects a stable low level of performance throughout.

Finally, for Inductive Reasoning (Figure 5.12) the subtypes indicate different combinations of ability level and age at which decline is first noted. Subtypes B and D reflect virtual stability until age 64 and decline thereafter; the former from a high level and the latter from a low level of adult performance. Subtypes A and C, on the other hand, represent individuals with an earlier onset of decline; the former with a less pronounced rate of decline from a higher level. Rate of decline is greatest for Subtype D (.017 *SD*), with a more moderate annual rate of about .010 *SD* for the remaining subtypes.

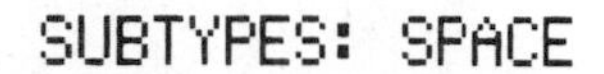

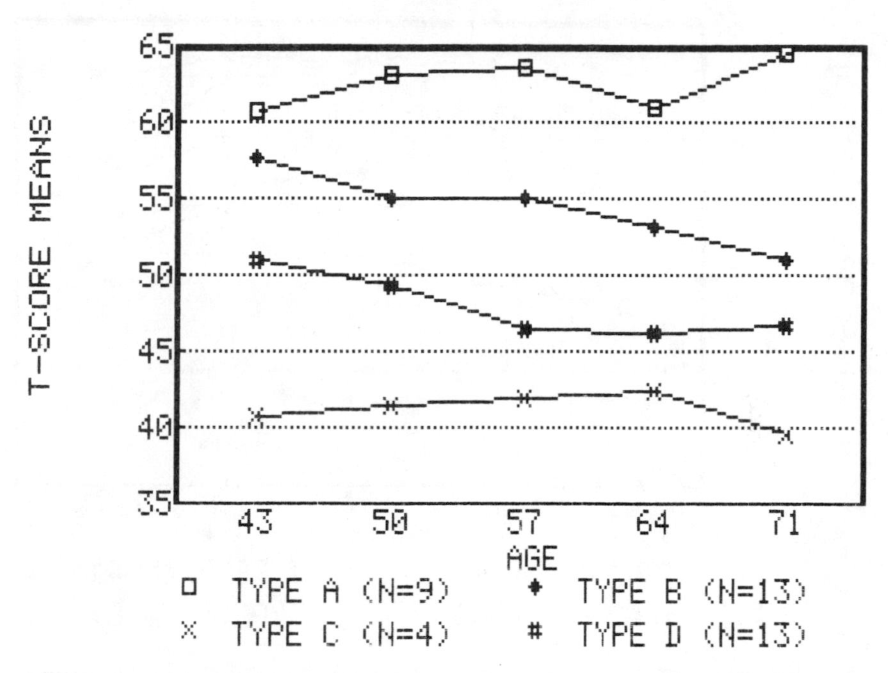

FIGURE 5.10 Subtype profiles for the young-old cohort on Verbal Meaning.

Individual Profiles

This presentation will be concluded by presenting profiles for four individuals on the measure of Verbal Meaning (recognition vocabulary). These profiles were chosen to illustrate several specific points. The first two profiles (Figure 5.13) represent two young-old women who throughout life functioned at very different levels. Subject 155510 is a high school graduate who has been a homemaker all of her adult life and whose husband is still alive and well-functioning. She started our testing program at a rather low level, but her performance has had a clear upward trend. The comparison participant subject (154503) had been professionally active as a teacher. Her performance remained fairly level and above the population average until her early sixties. Since that time she has been divorced and retired from her teaching job; her performance in 1984 dropped to an extremely low level, which may reflect her experiential losses but could also be a function of increasing

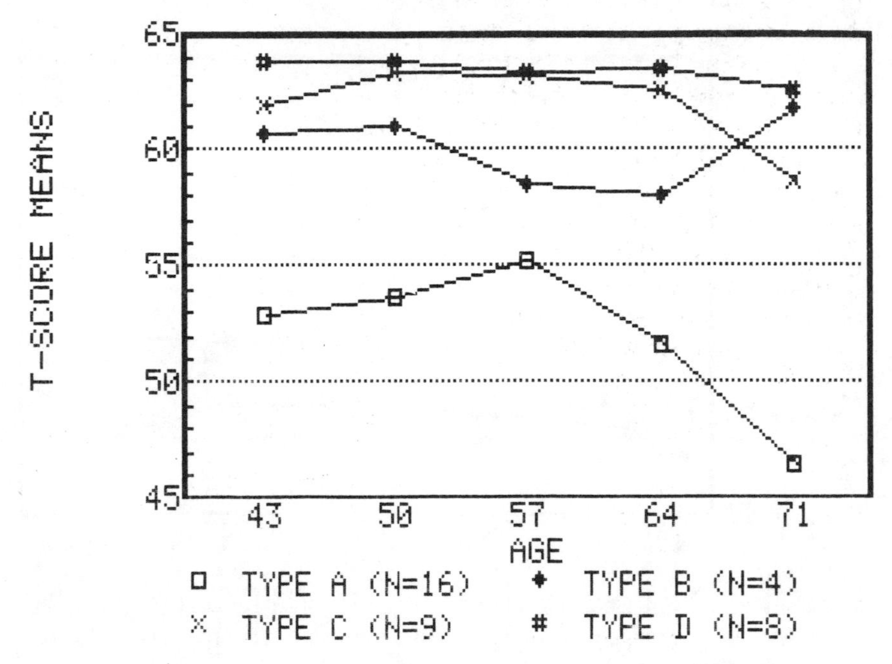

FIGURE 5.11 Subtype profiles for the young-old cohort on Spatial Orientation.

health problems (cf. Gribbin, Schaie, & Parham, 1980; Hertzog, Schaie, & Gribbin, 1978).

The second pair of profiles (Figure 5.14) shows the 28-year performance of two old-old men now in their eighties. Subject 153003, who started out somewhat below the population average, completed only grade school and worked as a purchasing agent prior to his retirement. He showed virtually stable performance until the late sixties; his performance actually increased after he retired, but he is beginning to experience health problems and has recently become a widower, and his latest assessment was below the earlier stable level. By contrast, subject 153013, a high school graduate who held mostly clerical types of jobs, showed gain until the early sixties and stability over the next assessment interval. By age 76, however, he showed substantial decrement that continued through the last assessment, which occurred less than a year prior to his death.

None of the four profiles just examined could have been directly predicted from our knowledge of group means, although each of these

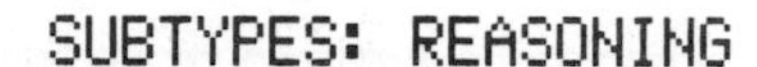

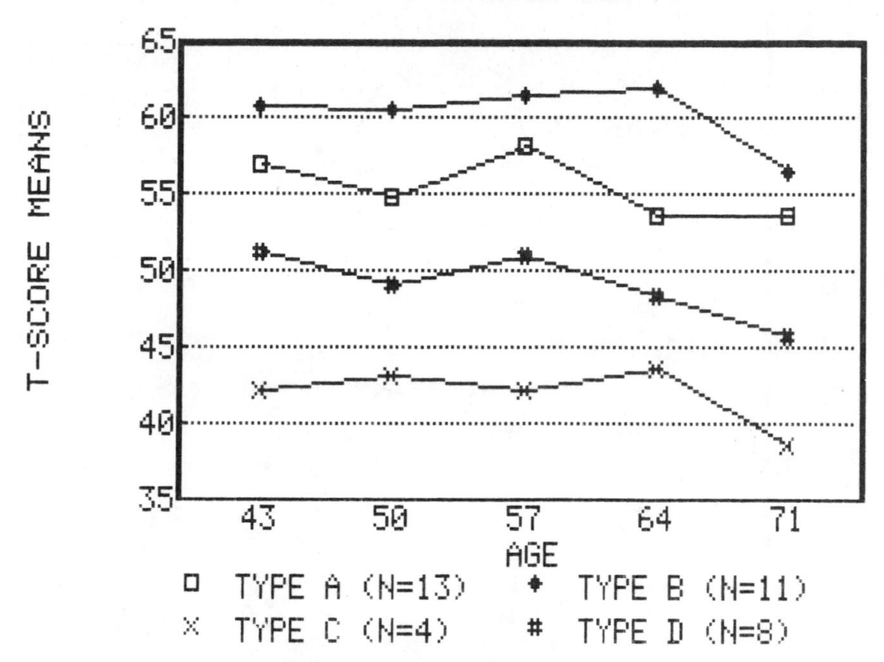

FIGURE 5.12 Subtype profiles for the young-old cohort on Inductive Reasoning.

profiles is quite similar to one of the subtypes exhibited earlier. Similarly, predictions of age changes for these individuals from the group-based estimates or rate of change would have been quite misleading.

SOME CONCLUDING THOUGHTS

In this chapter I have tried to lead the reader systematically from a representation of a data set that reflects conventional comparisons of average within-group age changes through a progressive disaggregation of such data. The purpose of this disaggregation was to reflect different levels of analysis selected in a manner that would bring us as close as possible to a reasonable representation of aging phenomena that occur in identifiable individuals, rather than remaining statistical abstractions. The data on individual differences could have been made even more dramatic had I relied only on cross-sectional data. But that would have defeated the purpose, for the essential information about

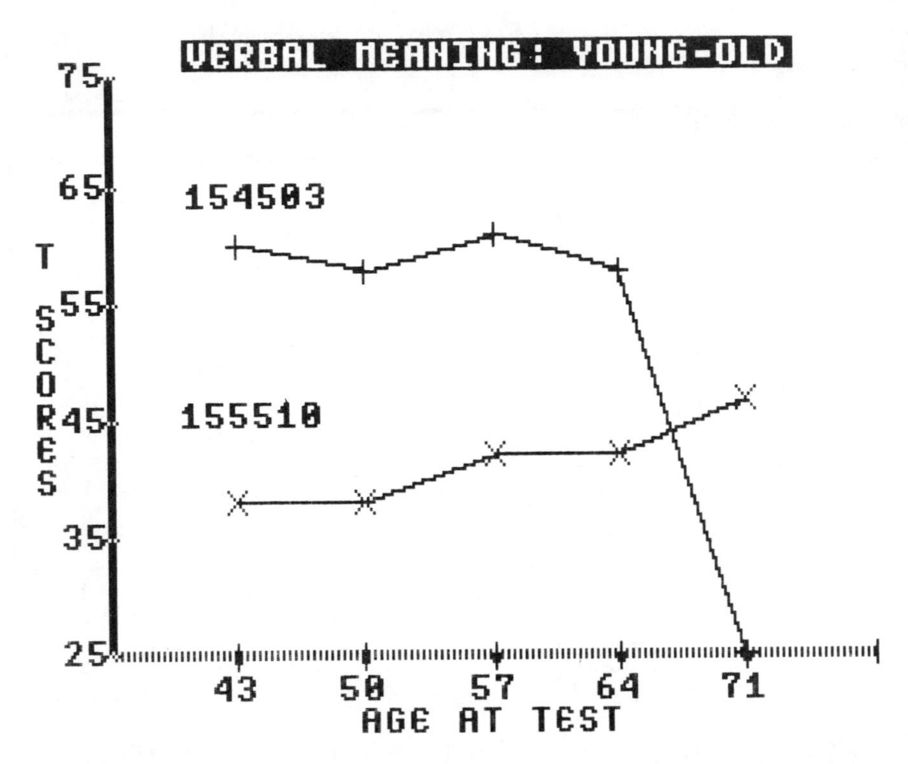

FIGURE 5.13 Individual profiles of Verbal Meaning for two young-old women.

individual aging can be elicited only from following the same persons as they age.

Once we begin to disaggregate our data by rate of change, it then becomes possible to address more meaningfully the question as to why some individuals age so much faster than others. Some hints are provided by contrasting individual profiles of subjects who experienced cognitive change either slowly or precipitously. It would go beyond the purpose of this chapter to delve more deeply into the variables that have been identified in previous work to distinguish those who age most rapidly from those whose behavior changes only minimally. Suffice it to indicate here that the most important variables identified thus far have been good cardiovascular health (Hertzog, Schaie, & Gribbin, 1978), active and involved life-styles (Gribbin, Schaie, & Parham, 1980), as well as flexible attitudes and intact interpersonal support systems (Schaie, 1984). In addition, there are obviously individual differences in the maintenance of energy level and of adequate sensory

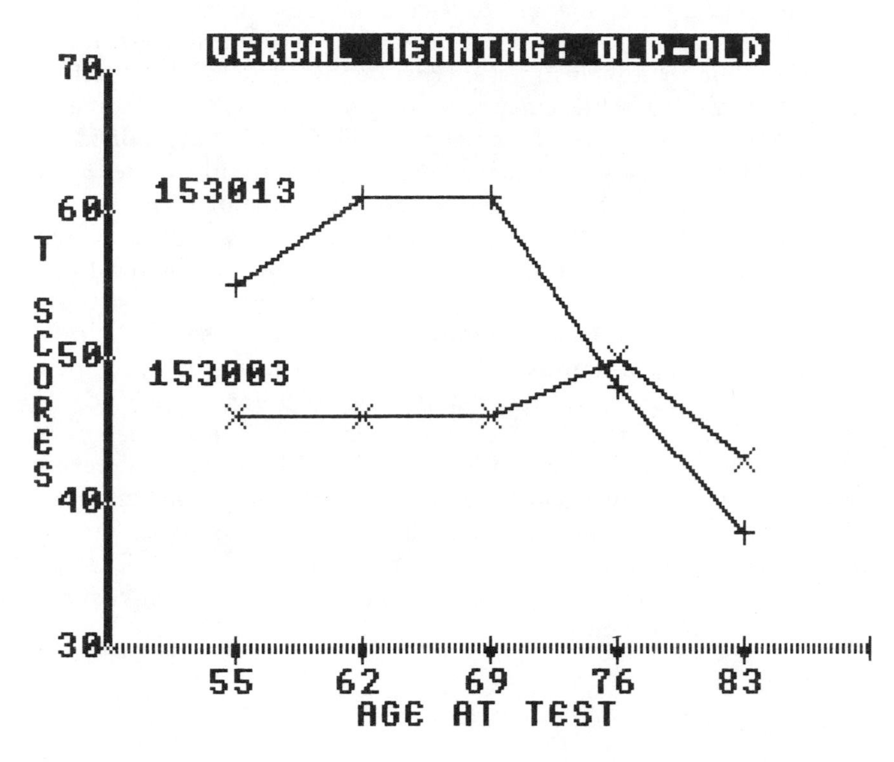

FIGURE 5.14 Individual profiles of Verbal Meaning for two old-old men.

functions that are related to adequate performance on the cognitive measures we have investigated (cf. Schaie, 1981).

There is, finally, a practical dilemma that remains to be addressed. All of the data presented here are, of course, retrospective in nature. If we wish to forecast future behavior of individuals or groups, we must use some model, or reference norms, that allow us to make such forecasts. Fortunately, there appear to be enough regularities in the data at hand to offer guidance as well as cautions. First of all, we might note that most of the identified subtypes reflect stable performance or increment into late middle age. It would therefore be reasonable to argue that the best predictor of accelerated aging may well be the detection of substantial decrement in the forties or early fifties. It is also important to note that accelerated aging appears only in a subset of those individuals who in midlife have functioned at a high level of performance. Apparently, rates of age change in the absence of pathology, at least until the seventies are reached, are quite small for those who are intellectually advantaged.

Complicating these matters further is the finding that there are individuals who show significant decrement at about the time they retire but who then plateau for another decade. The lowered performance of these individuals may reflect disuse and may thus be reversible (Schaie & Willis, 1986), or it may reflect the fact that the individual has established a new adaptation level consonant with the reduced demands of his or her personal environment (Schaie, 1977–1978).

We conclude then that rate of change in cognitive behavior as we age is a highly individuated phenomenon. Estimates derived from group norms reflect different rates for different abilities and suggest that there are differential cohort trends in rate as well as levels of performance. More importantly, strong arguments have been provided here to support the proposition that there is no uniform rate of cognitive aging that will adequately describe changes within individuals over time. Instead there are clearly differentiable patterns, a more detailed analysis of which now invites our attention, with the ultimate objective not only of understanding these differences but also of designing interventions that might make it possible to increase the proportion of the population that manage to maintain high levels of cognitive function to the end of their lives.

REFERENCES

Birren, J. E. (1959). Principles of research on aging. In J. E. Birren (Ed.), *Handbook of aging and the individual* (pp. 3–42). Chicago: University of Chicago Press.

Birren, J. E., Butler, R. N., Greenhouse, S. W., Sokoloff, L., & Yarrow, M. (1963). *Human aging*. Washington, DC: U.S. Government Printing Office.

Birren, J. E., & Cunningham, W. (1985). Research on the psychology of aging: Principles, concepts and theory. In J. E. Birren & K. W. Schaie (Eds.), *Handbook of the psychology of aging* (pp. 3–34). New York: Van Nostrand Reinhold.

Birren, J. E., & Morrison, D. F. (1961). Analysis of the WAIS subtests in relation to age and education. *Journal of Gerontology, 16*, 95–96.

Birren, J. E., & Renner, V. J. (1977). Research on the psychology of aging: Principles and experimentation. In J. E. Birren & K. W. Schaie (Eds.), *Handbook of the psychology of aging* (pp. 3–38). New York: Van Nostrand Reinhold.

Botwinick, J. (1977). Intellectual abilities. In J. E. Birren & K. W. Schaie (Eds.), *Handbook of the psychology of aging* (pp. 580–605). New York: Van Nostrand Reinhold.

Cohen, J. (1977). *Statistical power analysis for the behavioral sciences* (rev. ed.). New York: Academic Press.

Gribbin, K., Schaie, K. W., & Parham, I. A. (1980). Complexity of life style and maintenance of intellectual abilities. *Journal of Social Issues, 36,* 47–61.

Hertzog, C., Schaie, K. W., & Gribbin, K. (1978). Cardiovascular disease and changes in intellectual functioning from middle to old age. *Journal of Gerontology, 33,* 872–883.

Labouvie-Vief, G. (1985). Intelligence and cognition. In J. E. Birren & K. W. Schaie (Eds.), *Handbook of the psychology of aging* (2nd ed., pp. 500–530). New York: Van Nostrand Reinhold.

Schaie, K. W. (1977–1978). Toward a stage theory of adult cognitive development. *Aging and Human Development, 8,* 129–138.

Schaie, K. W. (1981). Psychological changes from midlife to early old age: Implications for the maintenance of mental health. *American Journal of Orthopsychiatry, 51,* 199–218.

Schaie, K. W. (1983a). The Seattle Longitudinal Study: A 21-year exploration of psychometric intelligence in adulthood. In K. W. Schaie (Ed.), *Longitudinal studies of adult psychological development* (pp. 64–135). New York: Guilford Press.

Schaie, K. W. (1983b). What can we learn from the longitudinal study of adult psychological development. In K. W. Schaie (Ed.), *Longitudinal studies of adult psychological development* (pp. 1–19). New York: Guilford Press.

Schaie, K. W. (1984). Midlife influences upon intellectual functioning in old age. *International Journal of Behavioral Development, 7,* 463–478.

Schaie, K. W. (1985). *Manual for the Schaie-Thurstone Test of Adult Mental Abilities (STAMAT).* Palo Alto, CA: Consulting Psychologists Press.

Schaie, K. W., & Hertzog, C. (1982). Longitudinal methods. In B. B. Wolman (Ed.), *Handbook of developmental psychology* (pp. 91–115). Englewood Cliffs, NJ: Prentice-Hall.

Schaie, K. W., & Willis, S. L. (1986). Can intellectual decline in the elderly be reversed? *Developmental Psychology, 22,* 323–332.

Spath, H. (1980). *Cluster analysis algorithms for data reduction and classification of objects.* New York: Wiley.

Thurstone, L. L., & Thurstone, T. G. (1941). *Factorial studies of intelligence.* Chicago: University of Chicago Press.

Thurstone, L. L., & Thurstone, T. G. (1949). *Manual for the SRA Primary Mental Abilities, 11–14.* Chicago: Science Research Associates.

Ward, J. H. (1963). Hierarchical grouping to optimize an objective function. *Journal of the American Statistical Association, 58,* 236–244.

Willis, S. L. (1985). Towards an educational psychology of the adult learner. In J. E. Birren & K. W. Schaie (Eds.), *Handbook of the psychology of aging* (2nd ed., pp. 818–847). New York: Van Nostrand Reinhold.

Willis, S. L., & Schaie, K. W. (1986). Practical intelligence in later adulthood. In R. J. Sternberg & R. K. Wagner (Eds.), *Practical intelligence: Origins of competence in the everyday world* (pp. 236–268). New York: Cambridge University Press.

6

A Factorial Analysis of the Age Dedifferentiation Hypothesis

Daniel F. Schmidt and Jack Botwinick

It was at the end of the 1940s when I (Jack Botwinick) first met James Birren. He was a section chief in what is now the National Institute on Aging Gerontology Center in Baltimore, and I was his newly arrived research assistant, fresh out of an M.A. program in psychology. We worked together a year or two and then each went our own ways, but not for long. We came together again in Bethesda in 1955 when he was chief of a larger section and I was one of his new Ph.D. staff members. Among Birren's early work was a 1952 study in factor analysis of the Wechsler-Bellevue test. His 1961 factor analytic study of the Wechsler Adult Intelligence Scale (WAIS) with Morrison remains a classic contribution. This present study is an outgrowth of those two studies.

The age-differentiation hypothesis (Garrett, 1946) states that as individuals approach adulthood, the organization of intelligence changes from a unified, general ability to a set of abilities or factors. Balinsky (1941) has suggested that, beyond early adulthood, dedifferen-

This study was supported in part by Training Grant AG-00030 and by Alzheimer Disease Research Center Grant AG 05681, both from The National Institute on Aging, National Institutes of Health.

tiation takes place with a reorganization of intelligence back to a single, general factor. Both the hypothesized initial differentiation and subsequent dedifferentiation have received mixed support (cf. Reinert, 1970).

Although Balinsky (1941) interpreted his factor analytic results as support for the hypothesis, other factor analytic studies of adult age differences in factor structure have reported three or four factors for all age groups (e.g., Berger, Bernstein, Klein, Cohen, & Lucas, 1964; Birren & Morrison, 1961; Cohen, 1957). The Birren–Morrison study was based on the WAIS standardization data and, like the others, reported factors with labels that are variants of General Intelligence, Verbal Comprehension, Perceptual Organization, and Memory. These studies, however, utilized older criteria for determining the number of factors in a data set. The present study utilizes a different criterion, one that has become popular with the now frequent use of computer-assisted analysis.

In a more recent study not directed to the dedifferentiation hypothesis but relevant to it nevertheless, Silverstein (1982) reported the results of a factor analysis of the Wechsler Adult Intelligence Scale–Revised (WAIS-R) standardization data. He reported a two-factor solution that was invariant among the age groups he studied. The two factors were interpreted as Verbal Comprehension and Perceptual Organization. Silverstein also reported results of a comparable analysis of WAIS standardization data. He found that the factor structure of the WAIS was highly similar to that of the WAIS-R, at least for the age groups studied. These results suggest that for the age range studied, 16 to 74, the organization of intelligence is invariant and characterized by specific Verbal and Performance factors.

The present study was also designed to examine the factor structure of the WAIS and the WAIS-R across the adult life span. The present study utilized the basic data examined by Silverstein (1982), but it also included additional WAIS data extending into late adulthood. The purposes of this study were to examine (1) the organization of intelligence across the adult life span in testing the dedifferentiation hypothesis, (2) the factorial similarity or difference between the WAIS and the WAIS-R, and (3) the adequacy of the various subtests as measures of Verbal and Performance components of intelligence. To these ends, age group differences in factor structure were examined but not in the manner of Silverstein, as will be explained later.

METHOD

The data consisted of intercorrelations between WAIS subtest scores and between WAIS-R subtest scores. The intercorrelations were those provided in the test manuals which included data of standardization samples of the different age groups. The WAIS data were reported by Wechsler (1955) and Doppelt and Wallace (1955). The WAIS-R data were reported by Wechsler (1981).

Each correlation matrix was analyzed by principal components analysis. The number of factors was determined by a criterion of eigenvalue greater than or equal to 1 (Kaiser, 1974). When more than one factor was found, they were rotated orthogonally to reach a final solution.

RESULTS

WAIS

The analysis of the WAIS data indicated a one-factor solution with the four youngest age groups: 18–19, 25–34, 45–54, and 60–64 years. The factor loadings of all subtests were high in each age group, indicating a General Intelligence factor for each age. This factor accounted for 58 to 61% of test score variance across the four age groups.

Two-factor solutions were found with the next two older groups (65–69 and 70–74). With both of these age groups, Factor 1 appeared as a Verbal factor with high loadings of Information, Digit Span, Vocabulary, Arithmetic, Comprehension, and Similarities subtests. It accounted for 37 and 42% of the variance for the 65–69 and 70–74 age groups, respectively. Factor 2 was a nonverbal, or Performance, factor with high loadings of Picture Completion, Picture Arrangement, Block Design, Object Assembly, and Digit Symbol subtests. Factor 2 accounted for 30 and 24% of variance for the 65–69 and 70–74 age groups, respectively. Picture Completion and Picture Arrangement were the only subtests that departed from this general pattern. Picture Completion loaded more highly on Factor 1 (.70) than on Factor 2 (.38) among those aged 70 to 74. Picture Arrangement had similar loadings on Factors 1 (.56) and 2 (.58) among those aged 65 to 69.

Finally, a three-factor solution was found for the oldest age group, 75 and over. Factor 1 was a Verbal factor, accounting for 33% of the variance, and Factor 2 was a Performance factor, accounting for 24% of the variance. Factor 3 was composed of only Similarities and Picture Arrangement subtests. This factor accounted for 18% of the variance.

WAIS-R

Analysis of WAIS-R data indicated one-factor solutions for ages 25–34, 35–44, 45–54, and 55–64. As with the WAIS, these factors appeared with each age group as a General Intelligence factor and accounted for 57 to 60% of the variance.

Two-factor solutions were indicated with the younger ages (16–17, 18–19, and 20–24) and also with the older ages (65–69 and 70–74). Factor 1 accounted for 36 to 40% of the variance and Factor 2 accounted for 20 to 27% of the variance. Although Factors 1 and 2 can be interpreted as Verbal and Performance factors, respectively, the pattern of their factor loadings differed more with age than did the WAIS factor loadings. Factor 1 showed high factor loadings of the verbal subtests with all age groups, but there was the major exception of the Digit Span subtest among 20- to 24-year-olds, where the loading was only .30. More important, Factor 1 showed substantial loadings of the traditional performance subtests, negating the label of "Verbal." For example, the Digit Symbol subtest loaded more highly on Factor 1 than on Factor 2 in age group 16–17, Picture Completion loaded more highly on Factor 1 in age group 20–24, and the Picture Arrangement subtest loaded more highly on Factor 1 in age groups 16–17, 18–19, and 20–24.

Factor 2 showed high loadings of most of the performance subtests, but there were exceptions just indicated. In addition, the Arithmetic subtest loaded more highly on Factor 2 (.61) than on Factor 1 (.47) among those aged 20 to 24, and the Picture Arrangement subtest showed similar loadings on Factor 1 (.51) and Factor 2 (.54) among 60- to 69-year-olds.

DISCUSSION

The present factor analytic results of the WAIS and the WAIS-R presented some support for the dedifferentiation hypothesis. Compared to those aged 16 to 24, for whom organization of intelligence was characterized by two ability factors, those aged 25 to 64 showed only one general ability factor, supporting the process of dedifferentiation. However, by late adulthood (65–75+), the organization of intelligence again showed the presence of two and even three specific ability factors. This suggests, in contrast to Balinsky's (1941) dedifferentiation hypothesis, that a single, general factor characterizes the organization of adult intelligence only for those aged 25 to 64. Among older adults the organization of intelligence was characterized by two or more specific ability factors. Thus, the process of dedifferentiation in early adulthood does not produce a lasting, final change in the structure of intelligence. In late adulthood, individuals appear to undergo one more change, one of differentiation.

These results and conclusions are not completely in accord with those of Silverstein (1982), not only because of the extended age range in the present study but also because of a different manner of determining the number of factors. Silverstein depended upon two different criteria, one an eigenvalue greater than or equal to 1 and the other a parallel analysis criterion (Humphreys & Ilgen, 1969). Based on the eigenvalue criterion, he found one or two factors, as in the present study; but based on the parallel analysis criterion, he found two or three factors. With these two different results, Silverstein concluded that an invariant two-factor solution best described the data across age groups. In the present study, the dedifferentiation hypothesis was tested with the eigenvalue criterion, traditional for such a purpose.

Some caution should be taken in drawing conclusions from these analyses regarding the dedifferentiation hypothesis. First, this is a cross-sectional study. A longitudinal study might yield different results: one study reported an invariant one-factor solution over a longitudinal 10-year period (Siegler & Botwinick, 1985). Second, neither version of the Wechsler test covers the universe of mental abilities. Third, although one or more factors appear consistently in these analyses, these factors together typically account for only 60 to 75% of the variance. Thus, 25 to 40% of the variance remains to be explained.

The factorial comparability of the WAIS and the WAIS-R can be assessed if it is assumed that the effects of time period (1955 vs. 1981) are not of any consequence. Factor analysis of the WAIS and the WAIS-R based on like-aged subjects yielded highly similar results. A General Intelligence factor accounted for the test scores in each age group that together ranged from 25 to 64 years. With those aged 65 to 74, two factors, Verbal Comprehension and Perceptual Organization, were found. Only with those aged 18 to 19 did the WAIS and the WAIS-R produce different factorial organization. A one-factor solution was indicated by their performance on the WAIS, while a two-factor solution was indicated by their WAIS-R performance.

The traditional grouping of Wechsler's subtests into Verbal and Performance tests was generally supported by the loadings of subtests on Verbal and Performance factors when two factors were indicated. Although some subtests loaded more highly on the opposing factor or loaded equally on the two factors, these exceptions usually occurred only with one age group. The Picture Arrangement subtest, however, showed inconsistency across age. Picture Arrangement loaded more highly on the Verbal factor with the youngest age groups taking the WAIS-R, but did not differentiate between Verbal and Performance factors with 65- to 69-year-olds on both the WAIS and the WAIS-R.

REFERENCES

Balinsky, B. (1941). An analysis of the mental factors of various age groups from nine to sixty. *Genetic Psychology Monographs, 23,* 191–234.

Berger, L., Bernstein, A., Klein, E., Cohen, J., & Lucas, G. (1964). Effects of aging and pathology on the factorial structure of intelligence. *Journal of Consulting Psychology, 28,* 199–207.

Birren, J. E., & Morrison, D. F. (1961). Analyses of WAIS subtests in relation to age and education. *Journal of Gerontology, 16,* 363–369.

Cohen, J. (1957). The factorial structure of the WAIS between early adulthood and old age. *Journal of Consulting Psychology, 21,* 283–290.

Doppelt, J. E., & Wallace, W. I. (1955). Standardization of the Wechsler Adult Intelligence Scale for older persons. *Journal of Abnormal and Social Psychology, 51,* 312–330.

Garrett, H. E. (1946). A developmental theory of intelligence. *American Psychologist, 1,* 372–378.

Humphreys, L. G., & Ilgen, D. R. (1969). Note on a criterion for the number of factors. *Educational and Psychological Measurement, 29,* 571–578.

Kaiser, H. F. (1974). An index of factorial simplicity. *Psychometrika, 39,* 31–36.

Reinert, G. (1970). Comparative factor analytic studies of intelligence throughout the human life-span. In L. R. Goulet & P. B. Baltes (Eds.), *Life-span developmental psychology* (pp. 467–484). New York: Academic Press.

Siegler, I. C., & Botwinick, J. (1985). Longitudinal comparisons of WAIS factor analyses. In E. P. Palmore, E. W. Busse, G. L. Maddox, J. B. Nowlin, & I. C. Siegler (Eds.), *Normal aging III.* Durham, NC: Duke University Press.

Silverstein, A. B. (1982). Factor structure of the Wechsler Adult Intelligence Scale-Revised. *Journal of Consulting and Clinical Psychology, 50,* 661–664.

Wechsler, D. (1955). *Manual for the Wechsler Adult Intelligence Scale.* New York: Psychological Corp.

Wechsler, D. (1981). *WAIS-R manual.* New York: Psychological Corp.

7

Intellectual Abilities, Speed of Response, and Aging

Walter R. Cunningham

In 1936, Irving Lorge published one of the earliest studies concerned with intelligence, speed of response, and aging. His idea was that speeded tasks may lead to underestimates of the capability of older people as compared with tasks that rely on level of performance of increasingly difficult questions. He studied this issue in a very simple way, administering three intelligence tests which varied speed demands. One was a very short, highly speeded test. Another, in contrast, had very liberal time limits, so that the score would reflect the person's ability to answer questions correctly without regard to speed of response. A third test was intermediate between the two. The three tests were administered to persons varying in age, and it was found that age differences were most pronounced for the highly speeded test and smallest for the least speeded test. Lorge interpreted this result as a performance deficit rather than a decline in capability.

THE DEVELOPMENT OF THE BIRREN HYPOTHESIS

James Birren developed a different way of looking at Lorge's result. Birren's graduate studies were interrupted by World War II, but after the war, in 1947, he received his Ph.D. and began a program of research on aging and behavior. Birren began doing research on various aspects of aging, particularly on perception and psychophysiology. Postdoctoral work with L. L. Thurstone encouraged an interest in intelligence and aging as well.

Birren had a good idea, an idea that has already had a profound effect on the psychology of aging and may have an even greater impact in the future. He noticed, as have most researchers who have studied cognitive aging, that older people performed a remarkably wide variety of tasks more slowly than younger people, on average. There was nothing particularly new in that finding. Decades before, Galton had shown older people to be slower (Koga & Morant, 1923). Even the ancient Greeks were aware of this (see Nestor's speech to the Greek host at the funeral games of Patroklos in the *Iliad*, p. 555 in the 1975 translation by FitzGerald).

Birren's good idea, which began to coalesce around 1948, was that perhaps these changes in speed could be used as an explanatory variable that would help account for changes in behavior as people aged. This approach was the reverse of the usual psychological thinking: Typically speed of response was thought of as a dependent variable, whereas Birren was suggesting that speed of response could be an independent, explanatory variable that would account for age changes in other variables.

The above idea formed the basis for dozens of research projects. Birren's early research interests were broad; spanning psychophysiology, perception, and intellectual functioning. His first papers in the 1940s addressed diverse issues; but as time went on, the theme of the relationship between aging, speed of response, and its influence on behavior began to predominate. From 1950 to 1965, he published over 30 empirical papers related to the slowing of behavior and aging. Birren and his colleagues studied the relationship between age and reaction time (Birren, 1955; Birren & Botwinick, 1955a,b; Birren, Cardon, & Philips, 1963; Birren, Riegel, & Morrison, 1962; Birren & Spieth, 1962; Botwinick & Birren, 1965), intellectual performance (Birren, 1952; Fox & Birren, 1949, 1950a,b) with a particular emphasis on the role of speed (Birren, Allen, & Landau, 1954; Birren & Botwinick, 1951a,b; Birren & Morrison, 1961), and psychophysiological reaction, perception, and other variables (Birren, 1947; Birren, Bick, & Fox, 1948; Botwinick, Brinley, & Birren, 1957; Kay & Birren, 1958; Riegel & Birren, 1965).

The insights obtained from this vigorous research program were presented in a paper entitled "Age changes in speed of behavior: Its central nature and physiological correlates," which was published in 1965 (Birren, 1965). To me, it is the most conceptually satisfying statement of Birren's ideas.

Birren wrote or coauthored over 60 reviews in the field of the psychology of aging and edited or coedited numerous volumes and handbooks. Apart from his own empirical and conceptual contributions, Birren had a tremendous influence on the field of gerontology through his tireless organizational work, both within psychology and across a variety of other disciplines. His activities were instrumental in influencing others to become active in aging research. He has trained many students.

BIRREN'S INFLUENCE ON MY WORK

As a graduate student, my own research activities grew fairly directly out of Birren's interest in speed and the psychometric orientation provided by his work with Thurstone at Chicago. When I went to graduate school, I was interested in the concept of intelligence. Under Birren's influence, I began looking at the relationship between aging and intellectual functioning.

My interest in speed and intelligence was sharpened considerably by the results of my master's thesis (Cunningham, 1974a; Cunningham & Birren, 1976). Birren encouraged me to carry out a longitudinal follow-up study, and I found significant declines in level of performance for a highly speeded Relations factor. The other intelligence variables studied in this effort showed marked differences across cohort, but the highly speeded Relations factor showed negligible cohort differences. This result is consistent in a general way with Birren's idea that changes in speed of response are a primary result of the aging process.

Birren encouraged me to do a dissertation in which I studied the structure of intelligence in relation to age, again using longitudinal data (Cunningham, 1974b). Research on the structure of intelligence involves analysis of the interrelationships between various measures of intellectual functioning; that is, test scores. Studying the structure of intelligence as a function of age involves examination of possible changes in interrelationships among measures of intellectual functioning and underlying hypothetical constructs often referred to as factors. This is usually studied with factor analytic methods.

In my dissertation, I found that there were consistent changes in structure that appeared to be most pronounced for highly speeded variables. This result was similar to earlier cross-sectional studies by Birren and others (e.g., Birren & Morrison, 1961) utilizing slightly different factor analytic models. The methods employed required independent (or orthogonal) factors. My results suggested that correlated (or oblique) factors might be more appropriate for the data, particularly at the age level of about 60 years. I therefore reanalyzed the data, using recently developed techniques of confirmatory factor analysis. I found that the intellectual abilities became more interdependent with age (Cunningham & Birren, 1980). However, because the number of tests and factors included in this study were quite limited, more extensive work was needed.

THE STRUCTURE OF INTELLECTUAL ABILITIES AND THE SPEED HYPOTHESIS

With these results in mind, I designed what has become known as the Florida Study, in which an extensive set of tests of intellectual ability was administered to over 1,000 senior citizens recruited in cooperation with the American Association of Retired Persons (AARP) in more than 30 cities throughout the state of Florida. The tests were organized in three separate test batteries: one battery consisted of highly speeded tests, a second allowed liberal time limits, and a third was intermediate. It was hypothesized that the largest changes in structure would be obtained for the highly speeded battery. Although I was not consciously aware of it at the time, both my hypothesis and my test groupings repeated in modern multivariate structural context the concepts of Irving Lorge (1936) from his study of almost 40 years ago, which predated Birren's career.

Even though I was studying differences in structure rather than differences in level (averages), the results were conceptually the same: age was most strongly associated with the highly speeded tests. It was found that the relationships between the tests and the factors were about the same for young and old, but the underlying factors were more interdependent in the old than in the young, thus replicating on a wider scale my dissertation results. This finding of greater interdependence has now been replicated in two other studies using slightly different factor analytic models (Hertzog & Schaie, 1986; Stricker & Rock, 1987).

My graduate students and I have now carried out a 7-year longitudinal follow-up on 450 elderly persons of the Florida Study. We have now replicated again in a longitudinal context the earlier cross-sectional finding that intellectual ability factors become more interdependent with age (Cunningham, Smook, & Tomer, 1985). We were also able to study changes in level in factor scores with age.

In 1976 and 1977, 1,116 subjects were recruited through the cooperation of the national and Florida officers of AARP in 29 cities throughout the state of Florida. They ranged in age from 51 to 89, with an educational range of 4 to 20 years. About 40% were male. The subjects were retested in 1983 and 1984. About 75% of the subjects rated their health as "good" or "excellent."

Although the study considered multiple indicators for ten separate factors of intellectual functioning, for the purpose of economy of presentation, only five will be discussed in this paper. The age-comparative construct validity of these factors has been extensively studied both cross-sectionally and longitudinally (e.g., Cunningham, 1980; Cunningham, Smook, & Tomer, 1985).

The ability factor definitions are as follows: *Expressional Fluency* is the ability to think rapidly of appropriate wording of ideas. A typical test for this factor requires the subject to construct a meaningful sentence from a set of letters which must be used as the initial letter in each word. For example, the letter set *a b c l* could be presented and a correct response would be "A bear cried loudly." *Figural perceptual speed* is speed in comparing figural material. In a typical task for this factor, a subject would be asked to quickly make comparisons between one target figure and four alternatives, one of which is identical to the target. *Symbolic perceptual speed* involves making comparisons and carrying out simple tasks involving visual perception with symbolic materials. An example would be to compare two strings of letters to evaluate whether or not they are identical. *Word fluency* is facility in producing isolated words that contain one or more structural, essentially phonetic restrictions, without reference to the meaning of the words. For example, the subject might be asked to write as many words as possible beginning with the letter F. *Verbal comprehension* is the ability to understand the English language. It is usually assessed with vocabulary tests (see Ekstrom, French, & Harman, 1976 for a discussion of these factors). Standardized scores were averaged across three or four test scores to estimate each factor score using integer weights (see Guilford, 1954, pp. 524–526; Guilford, 1965, p. 423; and Gorsuch, 1983, pp. 267–270, for a discussion).

Due to missing data, the number of subjects for each analysis varied slightly but was approximately 150 persons for each analysis. To evaluate cohort membership, the data were organized into four age groups with intervals of 7 years each. The median ages for the four groups at Time 1 were 59, 66, 73, and 80. Repeated measures analyses of variance were carried out for each factor with two between-subjects factors (age/cohort group and sex) and one within-subject factor (age/occasion of measurement).

Mean factor score plots for four cohorts across two occasions of measurement are portrayed in Figures 7.1 through 7.5. These figures portray data collapsed across sex because it was not a statistically significant variable. The first four factors showed statistically significant declines. The declines for expressional fluency and word fluency were complicated by pronounced cohort differences. The declines for symbolic perceptual speed and figural perceptual speed were consistently related to age and showed only trivial cohort differences.

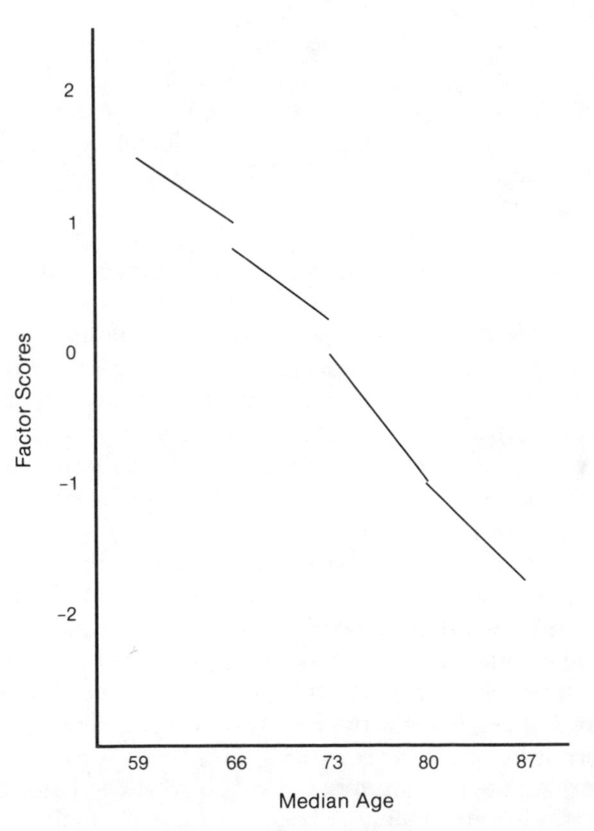

FIGURE 7.1 Age changes in symbolic perceptual speed.

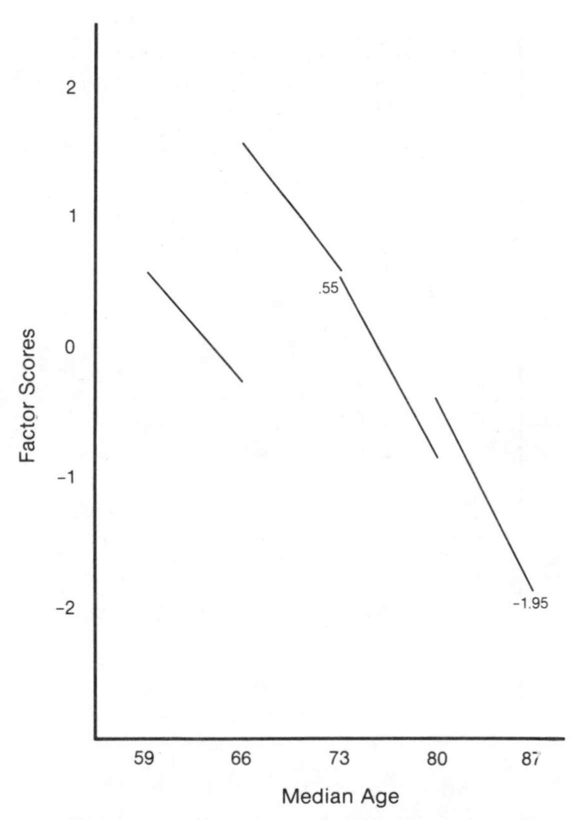

FIGURE 7.2 Age changes in word fluency.

The Verbal Comprehension factor shows a different pattern, however. Declines are very small and cohort differences are rather large relative to the age changes. This is an encouraging finding because this factor is probably a most important one in everyday life.

The finding of large age changes relative to cohort differences for the two perceptual speed measures is as hypothesized. In a general way, these results are consistent with others reviewed by Kausler (1982) supporting the generalization that simpler, more basic cognitive tasks may be influenced by cohort differences only to a minor extent. At the same time, the two fluency measures and verbal comprehension showed larger cohort differences and are similar to Schaie's results with the Primary Mental Abilities (e.g., Schaie, 1983; also see Schaie & Hertzog, 1983).

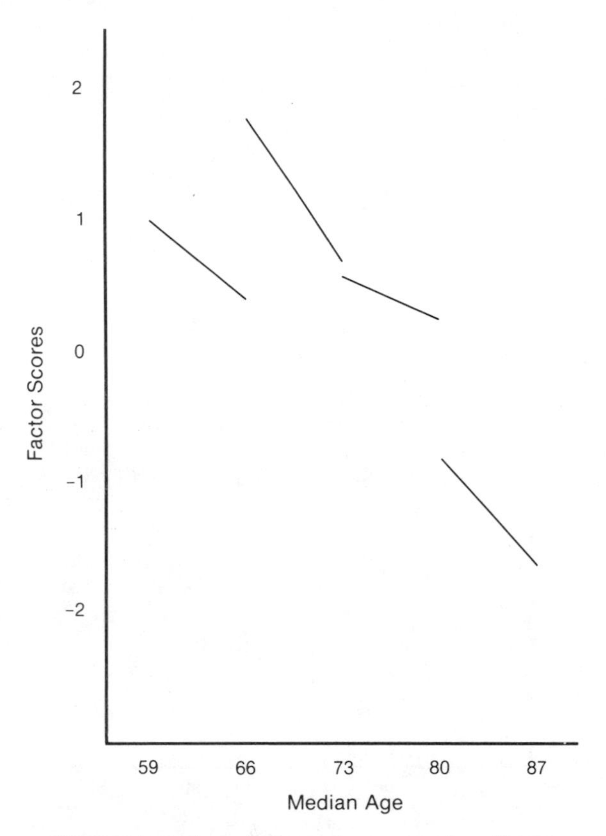

FIGURE 7.3 Age changes in expression fluency.

It is of some interest to speculate on the nature of cohort differences obtained. This is an important issue because there is no theory of cohort differences, and there is the possibility that empirically observed cohort differences may represent selection effects, retest effects, or other influences rather than reliable cultural differences, as is usually assumed (Birren, Cunningham, & Yamamoto, 1983). In the current results, the sample consists of retirees. For four of the five factors considered, the youngest cohort is performing on a lower level than the second youngest cohort. The same effect was apparent in another report (Cunningham, White, & Smook, 1985). It could be that early retirees differ from those who retire later in terms of health, motivation, or educational status. Thus, what appears to be a cohort effect at an empirical level may in fact represent specifiable selection factors. Of course, this line of reasoning is speculative, but the issue merits further study.

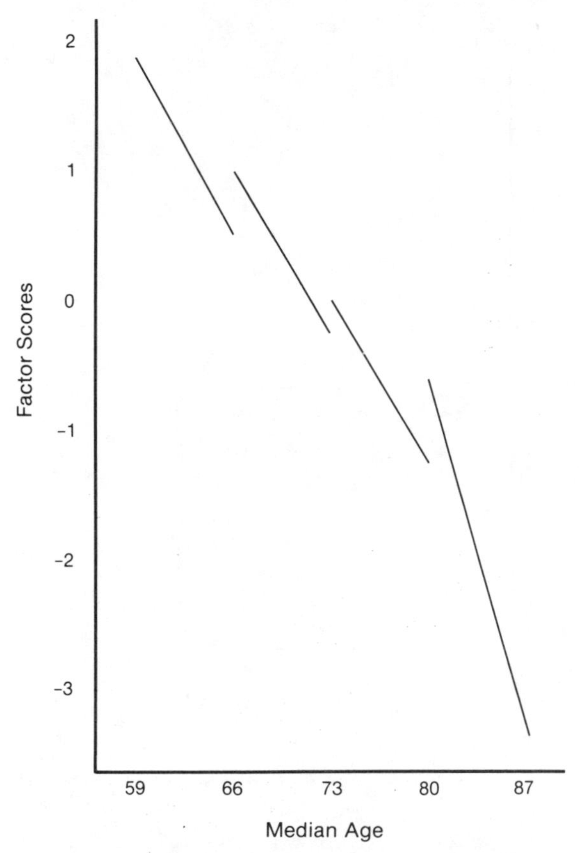

FIGURE 7.4 Age changes in figural perceptual speed.

While the results were very much in accord with the hypotheses, there was one surprise. It was expected that figural perceptual speed would show greater age declines and less cohort differences than symbolic perceptual speed. Although the former was found, the latter was not. It appears that symbolic perceptual speed may be less sensitive to cohort influences (cultural or otherwise) than figural perceptual speed.

Decline is very much in evidence in Figures 7.1 through 7.4, but it is important to recognize that these results are averages. Individual elderly persons do not necessarily decline at the same rate. Indeed, some older subjects decline to only a small extent even on the extremely age-sensitive perceptual speed tasks, while others show large declines on the verbal comprehension factor. It is obvious from examining individual scores that there are considerable individual differences in

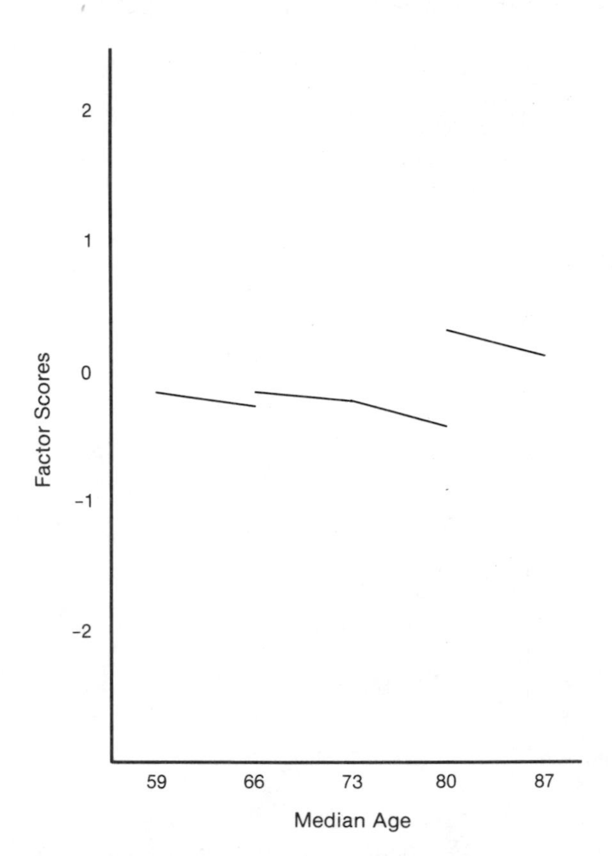

FIGURE 7.5 Age changes in verbal comprehension.

patterns of change that are far more complex than the relatively orderly relationships indicated by the mean plots. This is a very interesting area for future research. In particular, a major objective of future research would be to try to identify antecedents of patterns of age change.

In current work in progress, my graduate students and I are evaluating variables such as educational attainment and health ratings to predict longitudinal change in various intellectual abilities. We also intend to analyze the predictability of the perceptual speed factors with regard to longitudinal change with the data set described here and also with further longitudinal data currently being gathered. This latter activity stems directly from Birren's ideas about speed of performance as an explanatory variable to understand the nature of aging and behavior.

THE FUTURE OF THE BIRREN HYPOTHESIS

While much of this discussion has focused on my own research, it is important to recognize that there are now many researchers working on the relationship between aging and speed of response. When Birren began thinking about the speed of response and aging in the late 1940s, there were only a handful of researchers interested in this problem. Today there are scores of workers in this area producing important papers (e.g., Cerella, 1985; Salthouse, 1985). Also, it is worth noting that speed of response as an explanatory variable has seen a resurgence in mainline cognitive processing research (e.g., Hunt, Lunneborg, & Lewis, 1975; see also Chase, 1978) and also in intelligence research with regard to reaction time in children and young adults (Vernon, 1986).

It is clear from these developments in mainline cognitive research that there will be increased interest in further research on aging and speed of response. It is also clear that these efforts will continue to be influenced and enriched by Birren's ideas concerning the relationship between age and speed of response (e.g., Birren, 1965). For this, we are all in his debt.

TABLE 7.1 Test Instruments and Related Information

Test name[a]	Abbreviation	Factor[b]	Source[c]
Advanced Vocabulary	V5	V	ETS
Advanced Vocabulary 1	V41	V	ETS
Advanced Vocabulary 2	V42	V	ETS
Number Comparison	P2	P	ETS
Finding A's	P1	P	ETS
Letter Comparison	P4	P	VF
Word Arrangement	WA	Fe	G(ETS)
Simile Interpretation	SIM	Fe	G(ETS)
Expressional Fluency	EXF	Fe	G(ETS)
Prefixes I	PRE I	Fw	UF
Prefixes II	PRE II	Fw	UF
Suffixes	SUF	Fw	UF
Identical Pictures 1	P31	FiE	ETS
Identical Pictures 2	P32	FiE	ETS
Judgement of Size	JOS	FiE	G

[a] When a number follows a test name, it indicates that two scores were obtained from separately timed parts of the test.

[b] The factor abbreviations follow the ETS system (French, Ekstrom & Price, 1963): V, Verbal Comprehension; P, Symbolic Perceptual Speed; Fe, Expressional Fluency; Fw, World Fluency; FiE, Figural Perception Speed.

[c] ETS stands for Educational Testing Service manual, Kit of Referenced Tests for Cognitive Factors (French, Ekstrom, & Price, 1963). UF indicates an unpublished test developed at the University of Florida by the author, while G indicates a test developed in the Guilford Structure of Intellect Project and published by Sheridan Psychological Services, Inc., P.O. Box 6101, Orange, CA 91617. G(ETS) indicates a Guilford Test listed in the ETS manual.

REFERENCES

Birren, J. E. (1947). Vibratory sensitivity in the aged. *Journal of Gerontology, 2,* 267–268.

Birren, J. E. (1952). A factorial analysis of the Wechsler-Bellevue scale given to an elderly population. *Journal of Consulting Psychology, 16,* 339–405.

Birren, J. E. (1955). Age differences in startle reaction time of the rat to noise and electric shock. *Journal of Gerontology, 10,* 437–440.

Birren, J. E. (1965). Age changes in speed of behavior: Its central nature and physiological correlates. In A. T. Welford & J. E. Birren (Eds.), *Behavior, aging, and the nervous system* (pp. 1–26). Springfield, IL: Charles C. Thomas.

Birren, J. E., Allen, W. R., & Landau, H. B. (1954). The relation of problem length in simple addition to time required, probability of success and age. *Journal of Gerontology, 9,* 150–161.

Birren, J. E., Bick, M. W., & Fox, C. (1948). Age changes in the light threshold of the dark adapted eye. *Journal of Gerontology, 3,* 267–271.

Birren, J. E., & Botwinick, J. (1951a). Rate of addition as function of difficulty and age. *Psychometricka, 16,* 219–232.

Birren, J. E., & Botwinick, J. (1951b). The relation of writing speed to age and to the senile psychoses. *Journal of Consulting Psychology, 15,* 243–249.

Birren, J. E., & Botwinick, J. (1955a). Age differences in finger, jaw and foot reaction time to auditory stimuli. *Journal of Gerontology, 10,* 429–432.

Birren, J. E., & Botwinick, J. (1955b). Speed of response as a function of perceptual difficulty and age. *Journal of Gerontology, 10,* 433–436.

Birren, J. E., Cardon, P. V., & Philips, S. (1963). Reaction time as a function of the cardiac cycle in young adults. *Science, 140,* 195–196.

Birren, J. E., Cunningham, W. R., & Yamamoto, K. (1983). Psychology of adult development and aging. *Annual Review of Psychology, 34,* 543–575.

Birren, J. E., & Kay, H. (1958). Swimming speed in the albino rat: I. Age and sex differences. *Journal of Gerontology, 13,* 374–377.

Birren, J. E., & Morrison, D. F. (1961). Analysis of the WAIS subtests in relation to age and education. *Journal of Gerontology, 16,* 363–369.

Birren, J. E., Riegel, K. F., & Morrison, D. F. (1962). Age differences in response speed as a function of controlled variations of stimulus conditions: Evidence of a general speed factor. *Gerontologia, 5,* 1–18.

Birren, J. E., & Spieth, W. (1962). Age, response speed and cardiovascular functions. *Journal of Gerontology, 17,* 390–391.

Botwinick, J., & Birren, J. E. (1965). A follow-up study of card-sorting performance in elderly men. *Journal of Gerontology, 20,* 208–210.

Botwinick, J., Brinley, J. F., & Birren, J. E. (1957). Set in relation to age. *Journal of Gerontology, 12,* 300–305.

Cerella, J. (1985). Information processing rates in the elderly. *Psychological Bulletin, 98,* 67–83.

Chase, W. G. (1978). Elementary information processes. In K. W. Estes (Ed.),

Handbook of learning and cognitive processes (Vol. 5, pp. 19–90). Hillsdale, NJ: Erlbaum.

Cunningham, W. R. (1974a). *Age changes in human abilities: A 28 year longitudinal study.* Unpublished master's thesis, University of Southern California, Los Angeles.

Cunningham, W. R. (1974b). *Age changes in the factor structure of human abilities.* Unpublished doctoral dissertation, University of Southern California, Los Angeles.

Cunningham, W. R. (1980). Speed, age and qualitative differences in cognitive functioning. In L. W. Poon (Ed.), *Aging in the 1980's: Selected contemporary issues in the psychology of aging* (pp. 327–331). Washington, DC: American Psychological Association.

Cunningham, W. R., & Birren, J. E. (1976). Age changes in human abilities: A 28-year longitudinal study. *Developmental Psychology, 12,* 81–82.

Cunningham, W. R., & Birren, J. E. (1980). Age changes in the factor structure of intellectual abilities in adulthood and old age. *Educational and Psychological Measurement, 40,* 271–290.

Cunningham, W. R., Smook, G., & Tomer, A. (1985, October). *Age changes in factor structure of highly speeded tests.* Paper presented at the meeting of the Society for Multivariate Experimental Psychology, Berkeley, CA.

Cunningham, W. R., White, M., & Smook, G. (1985, August). *Longitudinal losses in intellectual abilities in the elderly.* Paper presented at the meeting of the American Psychological Association, Los Angeles.

Ekstrom, R. B., French, J., & Harman, M. (1976). *Manual for kit of factor-referenced tests.* Princeton, NJ: Educational Testing Service.

Fox, C., & Birren, J. E. (1949). Some factors affecting vocabulary size in later maturity: Age, education, and length of institutionalization. *Journal of Gerontology, 4,* 19–26.

Fox, C., & Birren, J. E. (1950a). The differential decline of subtest scores of the Wechsler-Bellevue intelligence scale in 60–69 year old individuals. *Journal of Genetic Psychology, 77,* 313–317.

Fox, C., & Birren, J. E. (1950b). Intellectual deterioration in the aged: Agreement between the Wechsler-Bellevue and the Babcock-Levy. *Journal of Consulting Psychology, 14,* 305–310.

French, J. W., Ekstrom, R. B., & Price, L. A. (1963). *Manual for kit of reference tests for cognitive factors.* Princeton, NJ: Educational Testing Service.

Gorsuch, R. L. (1983). *Factor analysis.* Hillsdale, NJ: Erlbaum.

Guilford, J. P. (1954). *Psychometric methods.* New York: McGraw-Hill.

Guilford, J. P. (1965). *Fundamental statistics in psychology and education.* New York: McGraw-Hill.

Hertzog, C., & Schaie, K. W. (1986). Stability and change in adult intelligence: I. Analysis of longitudinal covariance structures. *Psychology and Aging, 1,* 159–171.

Homer. (1975). *The Iliad.* (R. FitzGerald Trans.). Garden City, NY: Anchor Press/Doubleday. (Original work published c. 700 B.C.)

Hunt, E., Lunneborg, C., & Lewis, J. (1975). What does it mean to be high verbal? *Cognitive Psychology, 7,* 194–227.
Kausler, D. H. (1982). *Experimental psychology and human aging.* New York: Wiley.
Kay, H., & Birren, J. E. (1958). Swimming speed of the rat: Fatigue, practice and drug effects on age and sex differences. *Journal of Gerontology, 13,* 378–385.
Koga, Y., & Morant, G. M. (1923). On the degree of association between reaction times in the different senses. *Biometrika, 15,* 355-359.
Lorge, I. (1936). The influence of the test upon the nature of mental decline as a function of age. *Journal of Educational Psychology, 27,* 100–110.
Riegel, K. F., & Birren, J. E. (1965). Age differences in associative behavior. *Journal of Gerontology, 20,* 125–130.
Salthouse, T. A. (1985). Speed of behavior and the implications for cognition. In J. E. Birren & K. W. Schaie (Eds.), *The handbook of the psychology of aging,* (pp. 400–426). New York: Van Nostrand Reinhold.
Schaie, K. W. (1983). The Seattle Longitudinal Study: A 21-year exploration of psychometric intelligence in adulthood. In K. W. Schaie (Ed.), *Longitudinal studies of adult psychological development,* (pp. 1–19). New York: Guilford Press.
Schaie, K. W., & Hertzog, C. (1983). Fourteen year cohort sequential analysis of adult intellectual development. *Developmental Psychology, 19,* 531–543.
Stricker, L. J., & Rock, D. A. (1987). Factor structure of the GRE general test in young and middle adulthood. *Developmental Psychology, 23,* 526–536.
Vernon, P. A. (1986). The g-loading of intelligence tests and their relationship with reaction times: A comment on Ruchalla et al. *Intelligence, 10,* 93–100.

8

Neurobiological Models of Learning, Memory, and Aging

Diana S. Woodruff-Pak

An aspect of James Birren's research which has received less focus in recent decades is the work he did with aging animals. As an experimental psychologist, Birren spent a great deal of time in the 1950s with rats. In Figure 8.1 he is pictured with his favorite research animal as he looked in his days at the National Institute of Mental Health. A great deal that we know about learning and memory comes from research on the laboratory rat, and Birren recognized early in his career the value of conducting parallel studies in humans and animals.

In the 1950s Birren was interested in identifying the locus of the slowing of response speed in older organisms. He conducted parallel studies in humans and rats to examine the hypothesis that nerve conduction velocity slows with age. The study involving human subjects was carried out with Jack Botwinick, and Birren and Botwinick (1955) reasoned that by comparing reaction time responses from the jaw, finger, and

This research was supported in part by National Institute on Aging Senior Fellowship Award #1 F33 AG053212 and grants from the American Federation for Aging Research (AFAR).

foot, they would be varying the length of the peripheral nerve pathway. In their words,

> The purpose was to determine if the elderly subjects showed a disproportionate slowing of foot responses compared with the finger and jaw as a test of the hypothesis that the slowing of reaction time with advancing age is correlated with path length of the peripheral nerves. (p. 431)

They found that the old subjects were slower with all three kinds of responses, but they were not disproportionately slower with foot reaction time. Hence, peripheral nerve conduction velocity could not account for the slowing.

Following up the inferential study of peripheral nerve conduction velocity in humans with whom invasive measurement techniques were not possible, Birren and Wall (1956) directly measured peripheral nerve conduction velocity in the sciatic nerve of rats of various ages. The results of the study with humans was confirmed in that there was not a major decrease in conduction velocity in older rats.

Birren's research has always been characterized by experimentation and hypothesis testing. In his address as a recipient of the 1968 Distinguished Scientific Contribution Award from the American Psychological

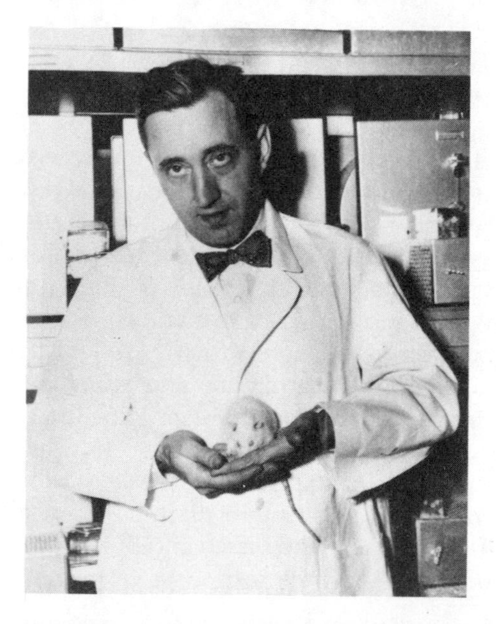

FIGURE 8.1 Photograph of James E. Birren and a research subject at the National Institute of Mental Health in Bethesda, Maryland, in the 1950s.

Association, he used the title, "Toward an Experimental Psychology of Aging." He stated, "The title of this article is meant to imply the desirability of an experimental psychology of aging as a goal toward which we can work" (Birren, 1970, p. 124). Later he explained that "we should encourage a descriptive and analytical psychology of aging while striving toward experimental control of the major causes of aging" (p. 135).

My recent research, summarized here, involving a model system approach to the study of learning, memory, and aging has been influenced substantially by my mentor, James Birren. Among the similarities in Birren's and my research are the use of animals in which invasive techniques can be utilized to measure the nervous system, an approach carrying out parallel research in humans and animals, and an experimental, hypothesis-testing orientation inherent in the model system approach.

CLASSICAL CONDITIONING AS A MODEL SYSTEM TO STUDY AGING

Classical conditioning of the eyelid closure response in the rabbit has been developed as a model system for analysis of the neurobiological substrates of learning and memory in the mammalian brain (Thompson et al., 1976). A substantial portion of the essential memory trace circuit underlying this basic form of associative learning has been defined (Thompson, 1983; Thompson, McCormick, & Lavond, 1986). Consequently, we are now in a position to identify and characterize the critical age-related changes in the brain that result in altered learning and memory performance as a function of age (Thompson & Woodruff-Pak, 1987; Woodruff-Pak & Thompson, 1985).

Two of the reasons highlighting the promise of eyelid conditioning in the rabbit as a model system for the study of learning and memory in aging follow: (1) age differences in the classically conditioned eyelid response are large, and (2) striking parallels exist between the age differences in eyelid conditioning in rabbits and humans. Behavioral and neurobiological studies of this simple form of learning in rabbits are likely to generalize to all mammals, including humans.

Importance of the Model System Approach

Ivan Petrovich Pavlov was the first to develop and use the model system approach to learning and memory. From the time he discovered the conditioned reflex, he saw it as a tool to investigate higher functions of the brain. Pavlov was also the first to observe age changes in classical conditioning. From his experiments on conditioning in dogs, Pavlov reported that old animals conditioned more slowly than young ones, and the responses of the old animals showed a different course of extinction. Birren (1961) acknowledged Pavlov's early contributions to the psychology of aging and favored his approach because of Pavlov's emphasis on the importance of the central nervous system in aging. Pavlov believed that the inhibition process was the first to succumb in old age. This hypothesis has continued to receive support in subsequent decades (Birren & Woodruff, 1983; Woodruff, 1985).

Karl Lashley, influenced by Pavlov, Bechterew, and Watson, was the first Western scientist to state explicitly the model system approach. In recent years the "model system" approach to analysis of the neuronal substrates of learning and memory has been valuable and productive. The basic notion is to utilize a preparation showing a clear form of associative learning in which neuronal analysis is possible. As the essential neural structures and pathways are defined, it becomes possible to localize and analyze cellular mechanisms underlying the memory trace (Cohen, 1980; Ito, 1982; Kandel & Spencer, 1968; Thompson, 1983; Tsukahara, 1981; Woody, Yarowsky, Owens, Black-Cleworth, & Crow, 1974).

Isadore Gormezano was the first to publish eyelid conditioning studies in the rabbit and introduced measurement of the nictitating membrane (NM) extension response. The rabbit is docile and tolerates restraint well, and the NM response is convenient to measure (Gormezano, Schneiderman, Deaux, & Fuentes, 1962; Schneiderman, Fuentes, & Gormezano, 1962). Consequently, an extensive amount of research has been carried out on classical conditioning of the rabbit NM response (Gormezano, 1972). Indeed, much of the general literature on classical conditioning is based on data collected in the rabbit NM paradigm. The NM and eyelid responses are highly correlated in rabbits (McCormick, Lavond, & Thompson, 1982), and the term *eyelid response* will be used to include studies in which either the NM or eyelid response was assessed.

One major advantage of the conditioned eyelid response is the fact that eyelid conditioning has become perhaps the most widely used paradigm for the study of basic properties of classical or Pavlovian conditioning of striated muscle responses in both humans and infrahuman subjects. It displays the same basic laws of learning in humans and other animals (Hilgard & Marquis, 1940). Consequently, it seems highly likely that neuronal mechanisms found to underlie conditioning of the eyelid response in rabbits will hold for all mammals, including humans. The conditioned eyelid response can be viewed as an instance of the general class of conditioned striated muscle responses learned with an aversive unconditioned stimulus (US) and the working assumption can be adopted that neuronal mechanisms underlying associative learning of the eyelid response will in fact be general for all aversive classical conditioning of discrete striated muscle responses.

Status of Knowledge Base on Brain Circuitry

Recent overviews of the current state of knowledge on neurophysiological substrates of classical conditioning in the rabbit have been presented (e.g., Thompson, McCormick, & Lavond, 1986). The data will be summarized briefly.

Hippocampus. The hippocampus and septohippocampal cholinergic system (system of neurons using acetylcholine as a neurotransmitter and reaching the hippocampus through the septum) is much involved in basic associative learning of the sort represented by eyelid conditioning in the rabbit. Its involvement is profound but modulatory in nature. Berger, Alger, and Thompson (1976) demonstrated that neuronal unit activity in the hippocampus increases markedly early in training and forms a predictive model in amplitude and time dimensions of the learned behavioral response. Hippocampal modeling occurs only under conditions where behavioral learning occurs. The response is generated largely by pyramidal neurons (Berger, Rinaldi, Weisz, & Thompson, 1983; Berger & Thompson, 1978b).

Although the hippocampus models acquisition of the classically conditioned response (CR), Solomon and Moore (1975) demonstrated that removing the entire hippocampus prior to training does not impair learning. An abnormally functioning hippocampus impairs learning, but the absence of a hippocampus does not. This would suggest that the hippocampus and septohippocampal system is exerting a modulatory action on the formation of the memory trace. The memory trace itself is not in the hippocampus, but the hippocampus can markedly influence the storage process. This result is somewhat parallel to the presumed role of the hippocampus in memory storage in humans (Squire, 1982).

In summary, the hippocampus appears to play a modulatory role in the acquisition of the classically conditioned NM/eyelid response. A model of the behavioral response develops in the hippocampus as learning proceeds. However, this neural activity is not the essential memory trace. Rabbits condition normally without the hippocampus. It is only when the hippocampus is functioning abnormally that basic learning is impaired, and facilitation in the hippocampus can accelerate learning. Thus, if aging processes result in abnormal functioning in the hippocampus, learning could be impaired. Interventions normalizing the old hippocampus could facilitate learning and memory.

Cerebellum. Animals with all brain tissue removed above the caudal level of the thalamus are able to learn the standard delay eyelid CR (Norman, Buchwald, & Villablanca, 1977). This result implies that the primary memory trace circuit is below the level of the thalamus. Systematic mapping of the entire midbrain, brain stem, and cerebellum by recording neuronal unit activity in trained animals indicates a substantial engagement of the cerebellar system in the generation of the conditioned response. Studies involving recording of neuronal unit activity from the deep cerebellar nuclei (especially the interpositus nucleus) over the course of training have in some locations revealed a striking pattern of learning-related growth in activity (McCormick, Clark, Lavond, & Thompson, 1982; McCormick & Thompson, 1984a,b).

Evidence from stimulating, recording, and lesion studies indicate that the ipsilateral interpositus nucleus is the site of the primary memory trace for the classically conditioned eyelid response. Furthermore, additional studies suggest that the memory trace for learning of all classically conditioned, discrete, adaptive somatic motor responses occurs in the cerebellum (e.g., Donegan, Lowry, & Thompson, 1983; Woodruff-Pak, Lavond, & Thompson, 1985; Yeo, Hardiman, & Glickstein, 1984).

A cerebellar model for learning. In the laboratory of Richard F. Thompson at Stanford University a hypothetical scheme or model of the neuronal system has been developed that could serve as the essential memory trace circuit for discrete, adaptive, learned somatic motor responses. This model is presented in Figure 8.2. The point of showing this seemingly complex model is to indicate the advanced state of understanding of the neural circuits underlying this simple form of learning.

To demonstrate the model, eyelid-closure and leg-flexion learning are used as the two behavioral examples. The site of the memory trace is assumed to be at the principal cells (labeled in the upper left as Purkinje cells). Principal cells are shown as Purkinje cells of the cerebellar cortex to demonstrate similarities with the models of Albus, Marr, Ito, and Eccles on cerebellar cortical function. The current data argue that the memory traces for the basic CRs are stored in the interpositus nucleus as well as in the cerebellar cortex, presumably by analogous circuitry (e.g., Woodruff-Pak, Lavond, Logan, Steinmetz, & Thompson, 1985).

In the model the conditioned stimulus (CS) "learning" input to a large number of principal cells comes from parallel fibers which have been

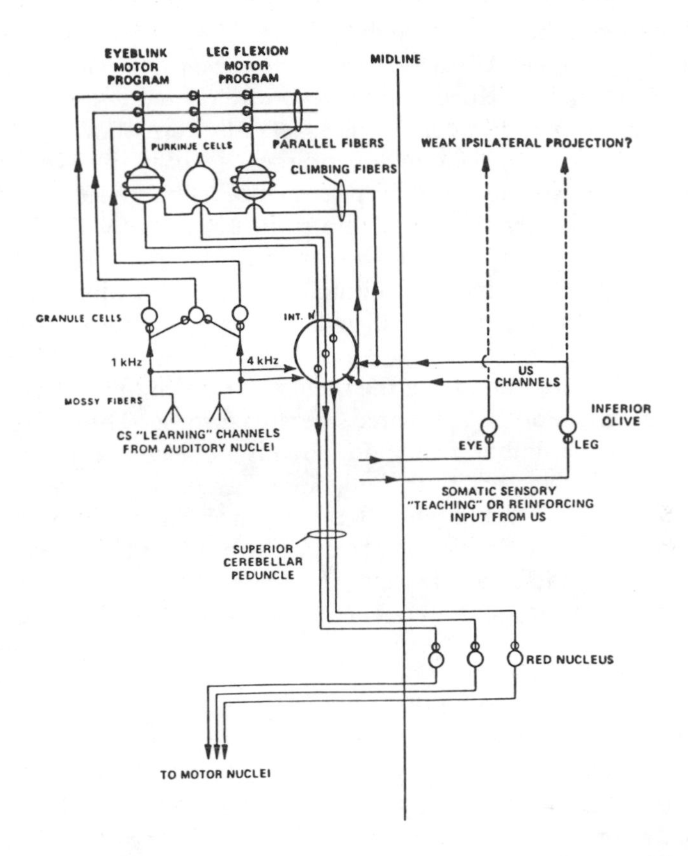

FIGURE 8.2 Scheme of hypothetical memory trace system for learning of discrete, adaptive somatic-motor responses to deal with aversive unconditioned stimuli. Interneurons are omitted. It is assumed that the site of the memory trace is at the principal neurons shown in the upper left under "motor programs" and/or at associated interneurons. The principal cells are labeled Purkinje cells of cerebellar cortex to show similarity with theories of cerebellar plasticity (Albus, Marr, Eccles, Ito). A given CS (1 kHz) activiates a subset of parallel fibers that in turn activate weakly all principal cells. The US pathway is assumed to be via the inferior olive and "climbing fibers." A given US is assumed to activate only a limited group of principal cells coding the motor program for the defensive response that is specific for the US (eyelid closure, leg flexion). When "parallel fiber" activation occurs at the appropriate time just prior to "climbing fiber" activation, the "connections" of the parallel fibers to the principal cells activated by the particular US are strengthened. Although this model is hypothetical, each aspect and assumption is amenable to experimental test (from Thompson, McCormick, & Lavond, 1986).

activated by mossy fibers. The US "teaching" input arrives at a specific and limited number of principal cells via climbing fibers from the inferior olive. A given US activates only specific principal cells coding the motor program for the defensive response related to the US (eyelid closure, leg flexion). The concept is essentially the following: (1) The principal cells for a specific motor program such as the eyeblink are activated (along with many other principal cells) by a neutral CS such as a tone; (2) shortly thereafter, the principal cells for the eyeblink response are selectively activated by climbing fibers (responding to the airpuff US); (3) with repeated pairing of the tone CS and airpuff US, the pathway involving the tone CS and the motor program for eyeblinking will be facilitated.

The basic assumption is that the memory trace circuit is "prewired" but the CS activation is "soft-wired" before learning. The connections from any CS to any motor program already exist before training, but they are too weak to elicit any behavioral responses before training. Activation of the CS circuit shortly prior to activation of the powerful motor-program-specific climbing fiber US input (the teaching or reinforcing input) results in a long-lasting facilitation of the previously weak "soft-wired" connections. The CS channel to the specific set of principal cells for a motor program becomes facilitated. This change is learning.

This scheme assumes that the essential output or efferent limb of the memory trace circuit is carried by the superior cerebellar peduncle, which crosses to relay in the magnocellular division of the contralateral red nucleus (Haley, Lavond, & Thompson, 1983; Madden, Haley, Barchas, & Thompson, 1983). From here the rubral (or red nucleus output) pathway crosses back and descends to act on motor nuclei and perhaps also on interneurons of the reflex pathways.

Of course, this theoretical scheme is tentative and hypothetical, but it does account for all of the available data. Each aspect and assumption of the scheme is amenable to experimental test. Indeed, since development of the scheme strong evidence has been amassed to support it.

The inferior olive is the source of climbing fibers to the cerebellum, while other afferents reach the cerebellum as mossy fibers. Evidence that the inferior olive-climbing fiber input is the essential teaching input for the learning of discrete, adaptive behavioral responses has been collected (McCormick, Steinmetz, & Thompson, 1985). It has also been recently demonstrated that stimulation of the rostromedial dorsal accessory olive can serve as the US. Stimulation of that area elicits an unconditioned response (UR), which in some cases is an eyeblink. Pairing a tone CS with the stimulation US resulted in the development of CRs within 100 to 200 trials. This is similar to the rate of acquisition using a corneal airpuff US (Mauk & Thompson, 1984).

Additional support for the model comes from studies using microstimulation of mossy fibers. Stimulation of the pontine nuclei or middle cerebellar peduncle demonstrated that mossy fiber stimulation can serve as an effective CS for classical conditioning (Steinmetz, Lavond, & Thompson, 1985a; Steinmetz, Rosen, Chapman, Lavond, & Thompson, 1986). Lesions of the middle cerebellar peduncle, the pathway for pontine afferents to the cerebellum, prevent conditioning (Solomon, Lewis, LoTurco, Steinmetz, & Thompson, 1986). Stimulation of the pontine nucleus as a CS and the dorsal accessory olive as a US mimics conditioning using a tone and airpuff (Steinmetz, Lavond, & Thompson, 1985b).

Evidence that the plasticity for learning occurs in the cerebellum rather than afferent to it comes from a recent study in which stimulation electrodes were placed symmetrically into the right and left dorsolateral pontine nuclei. Rapid transfer of training occurred from the left to the right stimulating electrode, indicating that the common area of cerebellar cortex stimulated by the symmetrical right and left dorsolateral pontine mossy fibers was involved as the site of plasticity (Steinmetz, Rosen, Woodruff-Pak, Lavond, & Thompson, 1986).

Finally, aging is known to affect acquisition and retention of the classically conditioned eyelid response in three species—humans, rabbits, and cats (Braun & Geiselhart, 1959; Gakkel & Zinina, 1953; Graves & Solomon, 1985; Harrison & Buchwald, 1983; Kimble & Pennypacker, 1963; Powell, Buchanan, & Hernandez, 1981; Woodruff-Pak, Lavond, Logan, & Thompson, 1987; Woodruff-Pak & Thompson, 1986). The model identifies sites in the cerebellum to test for age changes.

AGING AND CLASSICAL CONDITIONING OF THE EYELID RESPONSE

Research on Human Aging

The first investigators to report eyelid conditioning in older human subjects were Gakkel and Zinina (1953), who carried out their research in the U.S.S.R. The details of their work were described by Jerome (1959), who pointed out that the study was carried out in a home for invalids, with no young control group included. It was concluded that the process of conditioning was markedly prolonged in the aged invalids.

Braun and Geiselhart (1959) improved upon Gakkel and Zinina's design by selecting older subjects from a population residing in the community and by including comparison groups of young adults and children. There were dramatic age differences in the data. Braun and Geiselhart concluded that the main and striking result was the relative inability of the older subjects to acquire the conditioned eyeblink response.

Healthy community-residing old subjects with a mean age of 67 years were compared to young adults of a mean age of 20 in a careful eyeblink conditioning study by Kimble and Pennypacker (1963). Young subjects conditioned to a higher level than did old subjects. Age differences in the level of conditioning were statistically significant for the entire 60 trials and for the last 20 trials. By 60 trials of conditioning the young were presenting 50% CRs, while the old were presenting 30% CRs. Kimble and Pennypacker concluded that there is a difference between the conditionability of old and young.

We have recently reported eyelid conditioning data in the delay paradigm in adults ranging in age from 18 to 83 (Woodruff-Pak & Thompson, 1988b). Subjects were male and female university students and their parents, grandparents, and friends. They sat in an International Acoustics Company (IAC) chamber and wore safety goggles which held a minitorque potentiometer for the measurement of eyeblinks. Attached to the goggles was a tube resting 5 cm from the right cornea which delivered a 10-psi airpuff US of medical grade oxygen. A 1-kHz tone CS of 80-dB sound pressure level (SPL) was delivered to both ears through headphones. Timing and presentation of the stimuli and recording and analysis of the responses were controlled by a microprocessor programmed in Forth and machine language.

Subjects received 12 blocks of trials consisting of 1 unpaired and 8 paired tone and airpuff presentations. The 500-msec CS coterminated with a 100-msec US airpuff. The US onset was 400 msec after the CS onset. Subjects were given neutral instructions designed to neither facilitate nor to inhibit eyeblinks.

Results indicated that age differences in acquisition in classical conditioning of the eyelid response are large. A 4×12 repeated measures ANOVA comparing the effects of age and blocks of trials on percentage of CRs in the paired trials indicated highly significant age effects ($F = 8.81$; $p < .001$). Over blocks, percentage of CRs increased significantly ($F = 8.35$; $p < .001$). The interaction was not statistically significant. The correlation between total percentage of CRs and age for the entire group was $-.43$ ($p < .005$). These results are presented in Figures 8.3 and 8.4. Large age differences exist even in the relatively simple delay classical conditioning paradigm.

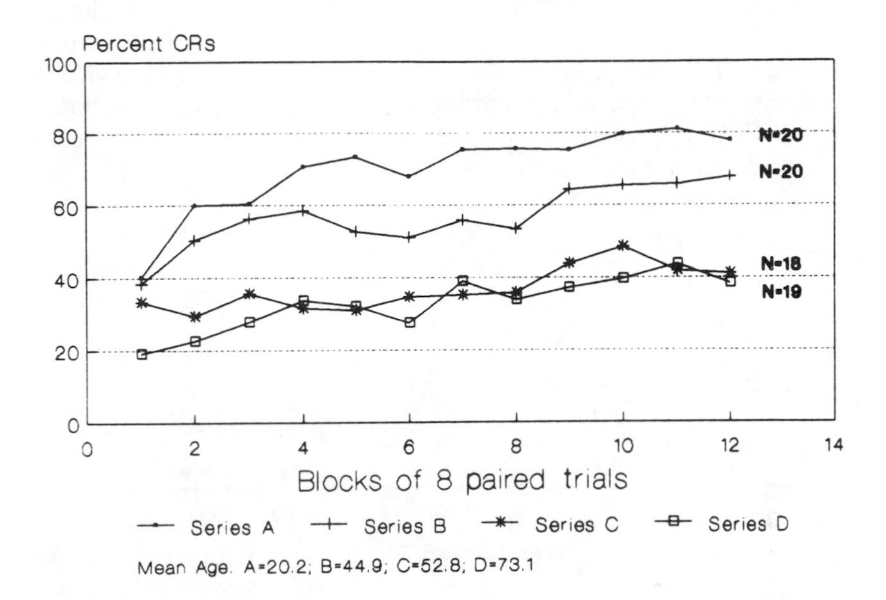

FIGURE 8.3 Acquisition in the delay classical conditioning paradigm of 77 men and women ranging in age from 18 to 83 years. The age range in the youngest group was 18–26 years, and for the older groups was 40–49 years, 50–59 years, and 61–83 years. Each block in this figure represents 8 paired trials of 500-msec, 80 dB SPL, 1-kHz tone CS coterminating with a 100-msec 10-psi airpuff US presented 400 msec after the onset of the CS (from Woodruff-Pak & Thompson, 1988b).

The trace classical conditioning paradigm provides a more difficult task for the subject because the CS and the US do not overlap. Figure 8.5 presents a comparison of the trace and delay paradigms. The term *trace* is used because the subject must form a memory trace of the CS to associate it with the later-arriving US. There are no classical conditioning data on the effects of aging in the human eyelid response in the trace paradigm, but our data on aging rabbits indicate that age differences in the trace paradigm in humans may be larger than in the delay paradigm. We have just initiated data collection in young, middle-aged, and aged humans in the trace conditioning paradigm.

Research on Animal Aging

The animal literature on aging of the classically conditioned eyelid response shows striking parallels to the behavioral aging data on

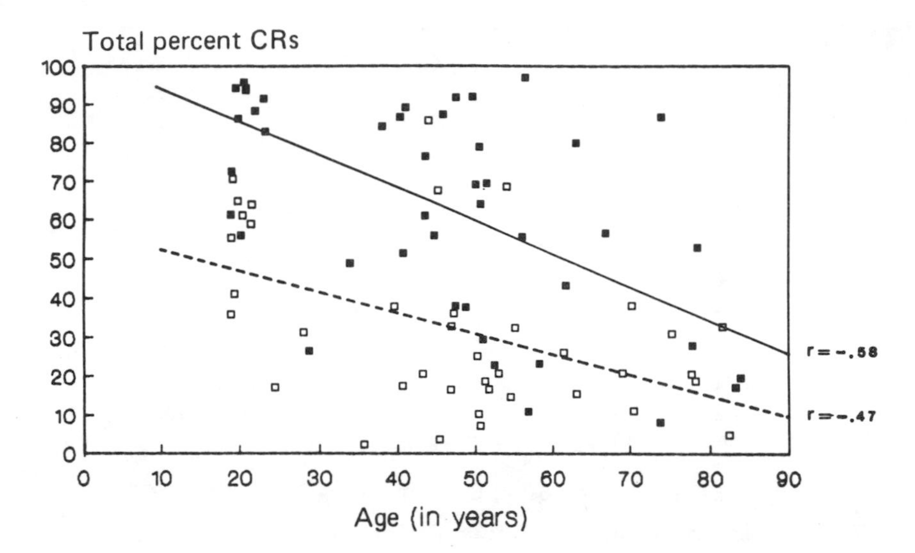

FIGURE 8.4 Relationship between age and total percentage of CRs for 96 paired CS–US trials in the delay classical conditioning paradigm for 83 human subjects ranging in age from 18 to 83. The data are from the same subjects as presented in Figure 8.3, and 6 subjects in the age range of 27–39 have been added. Subjects with unconditioned responses (URs) greater than 3.5 mm are shown as filled squares, and the correlation between age and percent CRs for this group of 44 subjects is −.58 ($p < .001$). Subjects with URs of 3.5 mm or less are shown as open squares. The correlation between total percentage of paired CRs and age was −.47 ($p < .005$) for these 39 subjects. Subjects with larger URs appeared to be more alert, and they conditioned more rapidly. There were no significant age differences in UR amplitude (from Woodruff-Pak & Thompson, 1988b).

humans. Using tone CS and shock US, Powell, Buchanan, and Hernandez (1981) found that 3- to 5-year-old rabbits (mean age, 40 months) required an average of 275 trials to the criterion of 10/10 CRs, which was significantly more trials than the average of 175 trials required by young rabbits (mean age, 6 months). In a subsequent experiment, Powell, Buchanan, and Hernandez (1984) trained 12 young and 12 old rabbits in a similar paradigm but with a CS+ and a CS−. Statistically significant interactions between the age *x* session and age *x* sex effects were found. Forty-month-old rabbits showed slower acquisition; older males learned more poorly than older females, while young males outperformed young females.

Using the delay classical conditioning paradigm in which a tone CS coterminated with a shock US, Graves and Solomon (1985) found no

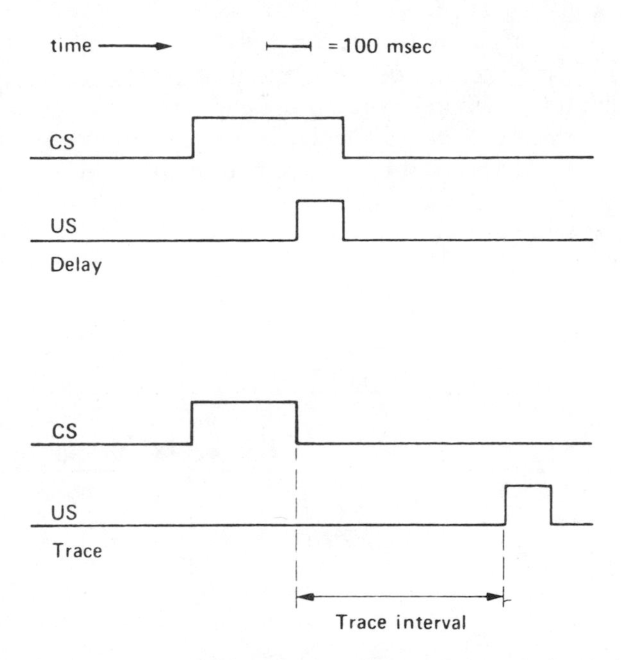

FIGURE 8.5 The standard delay (top) and trace (bottom) classical conditioning paradigms. CS, conditioned stimulus (in our laboratory typically a 1-kHz tone); US, unconditioned stimulus (shock or corneal airpuff). The trace interval was 500 msec (from Woodruff-Pak, Lavond, & Thompson, 1985).

age differences in trials to criterion for 6-month-old and 36- to 60-month-old rabbits. The task employed by Graves and Solomon may have been both easier and less aversive than that of Powell et al. (1981). In unpublished observations Berry and Thompson found that 5-year-old rabbits learned more slowly than young rabbits in the delay paradigm when an airpuff US was used. Taken together, these results indicate that seemingly small variations in the quality of the CS and US affect age differences in acquisition. These variations in parameters provide us with an opportunity to determine specific aspects in aging which result in the age differences in learning.

While Graves and Solomon (1985) found no age differences in classical conditioning in the delay paradigm, they did find large age differences in the trace paradigm. Eight 6-month-old and seven 36- to 60-month-old animals were conditioned in a trace paradigm. Trials to a criterion of 8/10 CRs were 461.00 for the young and 797.43 for the old animals. This difference was statistically highly significant.

With a corneal airpuff US and a 500-msec trace interval, Woodruff-Pak, Lavond, Logan, and Thompson (1987) found highly significant differences in conditioning between the three age groups of rabbits. These results are shown in Figure 8.6. Three-month-old rabbits attained criterion in a mean of 3.2 days (349 trials), while 30-month-old rabbits took a mean of 9.4 days (1,058 trials) to criterion, and 45-month-old rabbits took a mean of 11.75 days (1,392 trials) to criterion. These age differences in trials to criterion were statistically significant ($F(2,12) = 4.51$, $p < 0.05$).

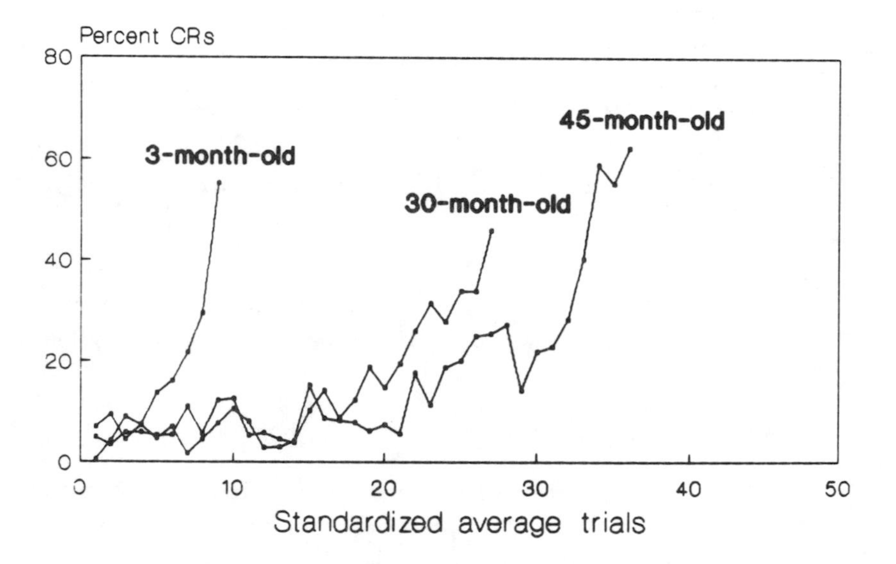

FIGURE 8.6 Acquisition in standard sessions for three age groups of New Zealand white rabbits of an average age of: 3 months (N=6), 30 months (N=5), and 45 months (N=4). Data are presented as standardized average trials because each rabbit attained criterion in a different number of trials. Mean trials to criterion for the young group was 349 trials (3 training days), for the middle group was 1058 trials (9 training days), and for the older group was 1392 trials (12 training days). To present the acquisition curve for each group, performance during each third of an average standard session is plotted. This means that the 3 training days for the young rabbits become 9 standardized average sessions, the 9 training days of the middle group become 27 standardized average sessions, and the 12 training days of the older group become 36 standardized average sessions (from Woodruff-Pak & Thompson, 1988a).

Variability between rabbits in trials to criterion also increased dramatically with age. This can be seen in Figure 8.7, which illustrates individual rabbit performance by age. While the youngest rabbits all attained criterion within a 4-day period, criterion performance in the 30-month-old rabbits ranged over 16 days; it ranged over 15 days for the 45-month-old group. These results suggest that aging of conditioning capacity occurs at different rates in different rabbits. Some of the 30- and 45-month-old rabbits performed as well as the 3-month-old rabbits, while others learned much more slowly. The correlation between age and training days to criterion was .70 ($p < .01$).

Age differences in trials to criterion in the delay paradigm run in sessions immediately after overtraining in the trace paradigm were not as great as age differences in the trace paradigm. The 3-month-old group attained criterion in delay conditioning in a mean of 101 trials. The 30-month-old group had a mean of 134 trials to criterion in delay, and the 45-month-old group had a mean criterion of 218 trials. These age differences in acquisition did not attain statistical significance.

In an effort to measure retention of the conditioning task in older rabbits, five of the older rabbits (mean age, 38 months) were retested on the delay task over an interval of 2 to 5 months. The correlation between trials to criterion in the initial learning of trace conditioning

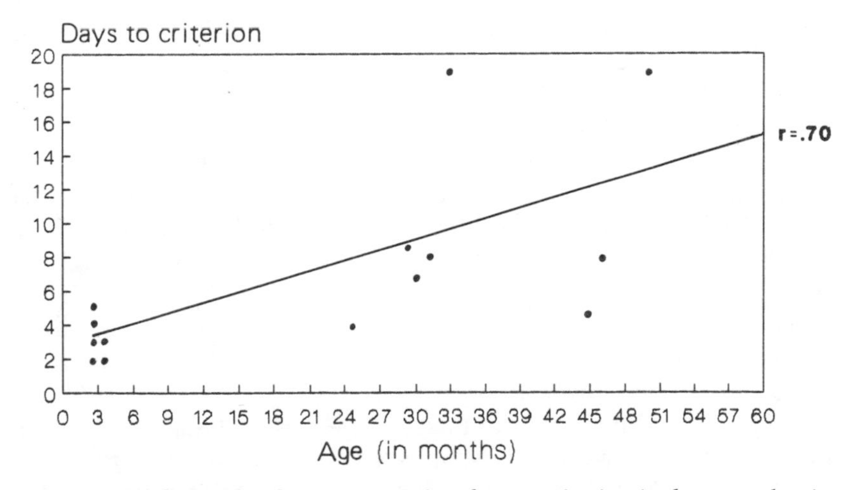

FIGURE 8.7 Relationship between training days to criterion in the trace classical conditioning paradigm for 15 rabbits between the ages of 3 and 50 months. The correlation between age (in months) and training days to criterion was .70 ($p < .01$) (from Woodruff-Pak, Lavond, Logan, & Thompson, 1987).

and the trials to criterion (corrected for the time interval between training and retraining) on delay conditioning was .99 ($p < .01$). The correlation between acquisition of the delay paradigm and retention of the delay paradigm 2 to 5 months later (corrected for the time interval between training and retraining) was .95 ($p < .01$). These high correlations indicate that retention of the task is strongly related to acquisition. The longer an older rabbit took to learn the trace and the delay conditioning tasks, the longer it took to relearn the delay conditioning task several months later. Young animals tested for retention in the delay paradigm several months after training needed little or no retraining to attain criterion on the task (Lavond & Thompson, 1984).

Large age differences in acquisition in the trace classical conditioning paradigm exist in rabbits as young as 2½ years. Age differences in the easier delay paradigm were much smaller and did not attain statistical significance in this study. However, acquisition was tested in the same rabbits after they were overtrained in the trace conditioning paradigm. It is possible that negative transfer from the trace conditioning task interacted with age and reduced the age differences. Thus, data from the present study cannot be used to resolve the discrepancy between Graves and Solomon (1985), who found no age difference in the delay paradigm, and Powell, Buchanan, and Hernandez (1981), who did find age differences in delay classical conditioning. Results from both of these laboratories on more difficult conditioning paradigms are supported by the present data, showing large age differences in trials to criterion in the trace paradigm.

The result that 2½-year-old rabbits take three times as many trials to attain criterion in the trace paradigm is surprising in that 2½ years is relatively young for a rabbit. The life expectancy of New Zealand white rabbits has not been empirically determined. Fox (1980) estimated that rabbit life expectancy is around 8 years. Female breeder rabbits are retired when the size of the litter of surviving young begins to decrease. This typically occurs between 18 and 24 months of age. Thus, aging as measured by reproductive capacity has declined by 2½ years, but the decline in learning capacity should not necessarily be linked to the decline in reproductive capacity.

All of the published studies of age differences in classical conditioning of rabbits involve comparisons of young virgin to retired breeder rabbits. While the age differences in classical conditioning are probably not entirely a function of comparing virgin to breeder rabbits, it is possible that frequent, repeated breeding accelerates aging processes. The three laboratories reporting age differences on the difficult conditioning tasks found them in rabbits as young as 3 years. Perhaps such age differences would appear later if the older rabbits had been bred less frequently.

While it is reasonable to suppose that frequent, repeated breeding might accelerate aging in female rabbits, it is less plausible to hypothesize that frequent mating accelerates aging in male rabbits. Indeed, sexual activity extends life expectancy in male rats (Drori & Folman, 1969, 1976). Powell et al. (1984) found the slowest acquisition of a difficult classical conditioning task in old male retired breeder rabbits, and the slowest learner in the trace conditioning paradigm in our study was an old male retired breeder. It appears unlikely that the differences in acquisition observed between young virgin and retired breeder rabbits can be attributed entirely to the effects of repeated breeding.

Another question raised by these data involves the age function over the rabbit life span of trials to criterion in classical conditioning. The oldest rabbit tested in this research was 50 months old at the beginning of the study. This rabbit took the longest to attain criterion. Fifty months is middle age for a rabbit. Does acquisition continue to take longer in older rabbits, or have we documented the greatest age differences in the 30- to 45-month-old range? There are no behavioral data for 6-year-old and older rabbits because old rabbits are almost impossible to obtain.

Classical conditioning of the eyelid response in cats shows aging effects remarkably similar to the effects reported in rabbits and humans. Harrison and Buchwald (1983) conditioned young and old cats in a relatively difficult paradigm in which the 4-kHz tone CS+ occurred 1,500 msec before the 50-msec shock US. The CS−, which the cats had to learn to ignore, were loud and soft clicks presented randomly throughout the session. Ten young cats, aged 1 to 3 years, met the criterion of 80% CRs in a mean of 270 trials. Nine of the 15 old cats, aged 10 to 23 years, failed to develop CRs at criterion level within 1,000 trials. The six old cats who developed criterion level CRs did so in a mean of 522 trials. Thus, old cats showed a marked deficit in a difficult classical conditioning paradigm. When old cats that had not demonstrated learning at the 1500-msec CS–US interval were trained at a 400-msec CS–US interval with no CS−, they attained criterion in a mean of 170 trials. These data, along with the trace conditioning data in rabbits, suggest that extending the time between CS and US onset amplifies age differences in learning.

RELATIONSHIPS BETWEEN AGING IN THE CENTRAL NERVOUS SYSTEM AND NEURAL MECHANISMS IN LEARNING AND MEMORY OF THE EYELID RESPONSE

A structure which has been identified as modulating acquisition of the classically conditioned eyelid response is the hippocampus. The primary memory trace for this response appears to be localized in the interpositus nucleus of the cerebellum. Thus, the model system approach can focus the study of aging mechanisms on the structures known to be involved in learning and memory processes in the classically conditioned eyelid response. In terms of deleterious effects of aging on acquisition and retention in eyelid conditioning, it should be possible to determine whether they are primarily on "processing" and modulatory phenomena (e.g., septohippocampal system), on the storage processes itself (cerebellar system), or on both.

Septohippocampal System

The human hippocampus is a structure affected most consistently and earliest by the histopathology of aging (Malamud, 1972), and aging changes have been documented in the hippocampus of the rat (Barnes, 1979; Landfield & Lynch, 1977) and the rabbit (Hernandez, Buchanan, Powell, & Shah, 1979). Tissue changes in the human brain, such as granulovacuolar degeneration of neurons, involve pyramidal cells of the hippocampus more than any other structure and are more observable in senile dementia and senile dementia of the Alzheimer's type (SDAT). Senile plaques are found most densely in the hippocampus, as are neurofibrillary tangles, which are most obvious in the pyramidal cells of the hippocampus and the third-layer pyramids of prefrontal and superior temporal neocortex (Scheibel & Scheibel, 1975). Tomlinson (1972) stressed that when neurofibrillary tangles are present in the apparently intellectually intact aged, they are found almost exclusively in the hippocampus. Berger and Thompson (1978a) found in rabbits that pyramidal cells are the critical elements in hippocampal neuronal plasticity during learning. Ironically, hippocampal pyramidal cells are the cells most affected in human aging.

Hippocampal unit correlates of learning suggested that the hippocampus models the NM response in old as well as in young animals (Woodruff-Pak et al., 1987). The hippocampal unit activity modeling of the NM response in a 4-year-old rabbit is shown in Figure 8.8. While unit modeling occurs in older rabbits, it takes longer to develop. This

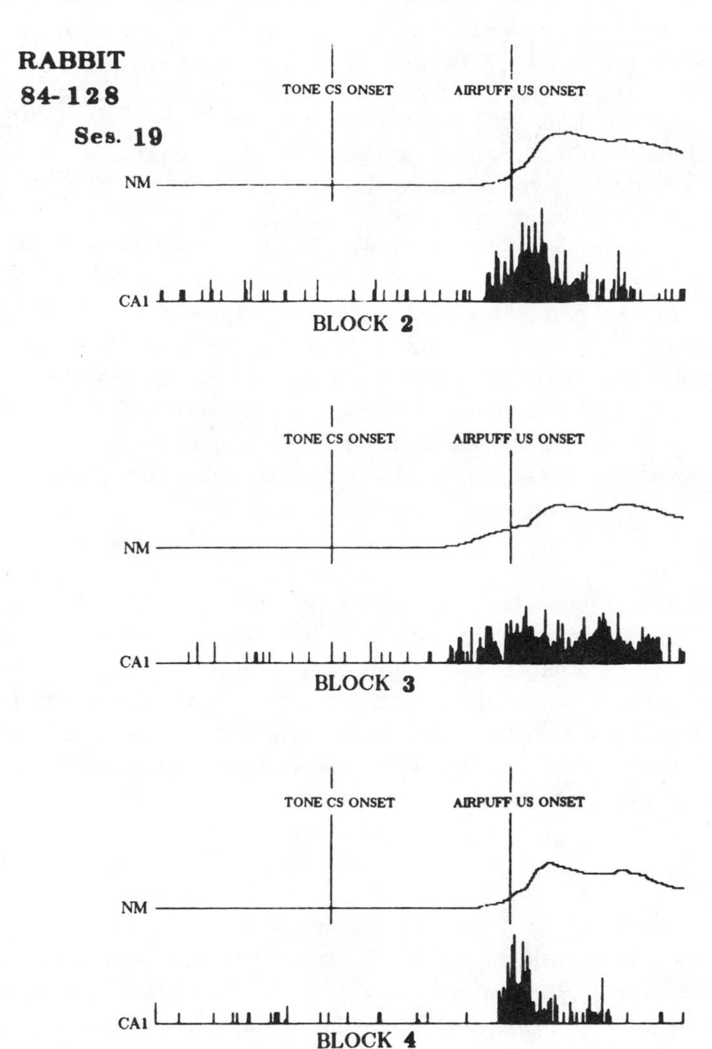

FIGURE 8.8 Examples of hippocampal unit activity modeling the conditioned nictitating membrane (NM) response in a 45-month-old New Zealand white rabbit. Each average represents 9 trials presented in the delay classical conditioning paradigm. The NM response shown in the top tracing of each of the three averages is an average of 9 trials. The hippocampal unit response shown below is a summation of activity in those same 9 trials. The first vertical line in each of the three averages is the onset of the tone CS, while the second vertical line represents the onset of the corneal airpuff US. The total time period represented by each average is 750 msec (from Woodruff-Pak, Lavond, Logan, & Thompson, 1987).

slower development of the hippocampal unit modeling parallels the slower acquisition of the CR. By the second session of training in the trace classical conditioning paradigm, when some 3-month-old rabbits have attained learning criterion, unit activity has increased significantly in the US period for young more than for old rabbits.

The age differences in hippocampal unit activity were in relation to the latency of onset of hippocampal modeling of the behavioral response. The hippocampal unit data in old rabbits was similar to unit data in young rabbits when the learning criterion was attained. Hippocampal unit activity in older rabbits models the CR. However, the conditioned behavioral response and hippocampal unit modeling of the behavioral response take much longer to develop in old rabbits. The hippocampal unit data suggest that age differences in the hippocampus may not account for age differences in acquisition of the CR. Hippocampal unit activity does model the behavioral eyelid response, even though the behavioral acquisition and hippocampal modeling are delayed in older rabbits.

These results suggest that it is important to focus on other structures in the aging brain, particularly the cerebellum. The cerebellum has been identified as essential for the retention of the eyelid response classically conditioned in the delay and trace paradigms (McCormick & Thompson, 1984a; Woodruff-Pak, Lavond, & Thompson, 1985). Neurophysiological age differences in the cerebellum may account for some of the age differences in learning.

Cerebellar System

The discussion on aging in the cerebellum is based on studies in humans, monkeys, and rats. Some years ago Harms (1944) reported that in the human cerebellum up to 25% of the Purkinje cells were lost in very old patients. Ellis (1920) reported greater loss in the anterior lobe of the cerebellum than in the hemispheres. Assessing Purkinje cell electrophysiology in Sprague-Dawley rats, Rogers, Silver, Shoemaker, and Bloom (1980) found a number of cell-firing parameters which were affected by age. In particular, increasing numbers of aberrant, very slow firing cells were encountered in older animals. Nandy (1981) reported a 44% decrease in the number of Purkinje cells in the left cerebellar cortex of 20-year-old rhesus monkeys as compared to 4-year-old monkeys. The number of granule cells in the same area were relatively equal in the two age groups. Purkinje neurons from old rats were significantly less sensitive to locally applied neurotransmitters than were neurons from young rats (Marwaha, Hoffer, & Freedman, 1981). Marwaha et al. hypothesized that there was a senescent postsynaptic change in noradrenergic transmission in Purkinje cells.

Consistent with the electrophysiologic and biochemical data showing age pathology of the cerebellar Purkinje cell are the anatomical changes in old rats reported by Rogers, Zornetzer, Bloom, and Mervis (1984). In Golgi-Kopsch sections, many 26-month-old Purkinje cells appeared defoliated, with small distal dendrites and spiny branchlets being the most affected. There was a significant decrease in the mean Purkinje cell area between 6-month-old rats and 26-month-old rats. The author suggested that the morphologic changes might be the hallmark of dying cells. In every vermis lobule examined, there was a significant senescent decrease in Purkinje neuron density. The mean number of Purkinje cells per millimeter of Purkinje cell layer declined from 16.6 cells/mm in young rats to 12.5 cells/mm in old rats. Related to the Purkinje cell loss was a loss in synaptic density.

The age change in the cerebellum which has been reported most frequently is a decline in Purkinje cells. Purkinje cells are the output neurons of the cerebellar cortex. The output cells of the cerebellar nuclei are, of course, by definition the principal cells. Our model, presented in Figure 8.2, suggests that an important focus of aging research should be on the Purkinje and principal cells. If they show changes similar to age changes in Purkinje cells in other species, this would provide evidence for age-related decline in the central component of the memory trace. If there is defoliation of the synaptic tree or the total loss of Purkinje and/or principal cells in aging, the net effect would be to eliminate the central coordination between the climbing-fiber "teaching" input and the mossy "parallel"-fiber "learning" input.

The result of Purkinje and principal cell elimination in this model would be the absence of the memory trace. Reducing the number of Purkinje and principal cells might reduce the conditionability of the organism—a result comparable to the behavioral data on classical conditioning of the eyelid response presented above. Indeed, Braun and Geiselhart (1959) and Kimble and Pennypacker (1963) used the terms "relative unconditionability" and "difference between the conditionability of old and young subjects," respectively, to describe their results with aging human subjects.

With regard to the two major input systems in the hypothetical schematic for classical conditioning of discrete adaptive responses—the climbing-fiber input from the olivocerebellar system providing US teaching information, and the parallel fiber input from the sensory systems providing CS learning information—data from the rat suggest that aging deficits may be limited to the parallel fibers. Analysis of the excitatory projection of single climbing-fiber efferents from the inferior olive making multiple synaptic contacts onto apical dendrite shafts of single Purkinje neurons revealed no apparent change in the number of climbing-fiber-mediated bursts or climbing-fiber spikes in rats from 3 to 28 months of age (Rogers et al., 1980). However, measures of parallel-fiber conduction velocity, refractory period, threshold, and current dependent volley amplitude in rats aged 5 to 7 and 24 to 26 months indicated senescent changes in the parallel fiber system (Rogers, Zornetzer, & Bloom, 1981). From the perspective of the model, this would mean that the CS input is less efficient in old organisms.

Our initial explorations of the aged rabbit cerebellum support the hypothesis that the CS input channels are less efficient (Woodruff-Pak, Steinmetz, & Thompson, 1986). Age differences in peripheral portions of the auditory CS pathway could account for the observed age differences in conditioning in rabbits. In this study we bypassed peripheral portions of the CS pathway and input the CS more centrally by electrical microstimulation of mossy fibers from the dorsolateral pontine nucleus.

Seven rabbits, ranging in age from 30 months to 44 months (mean age, 36 months), were anesthetized and implanted bilaterally with bipolar stimulating electrodes in the dorsolateral pontine nucleus or mossy fibers. Training consisted of 108 trials per day in which stimulation near the dorsolateral pontine nucleus served as the CS. The US was corneal airpuff. Rabbits were trained to a criterion of 8/9 CRs. Six of the rabbits were then trained with a tone CS to determine if the mossy-fiber stimulation CS would transfer to an auditory CS.

Results indicated that older rabbits could be trained with dorsolateral pontine stimulation as the CS, but they take many more trials to criterion than 3-month-old rabbits. The mean number of trials to learning criterion was 529.6. This is over five times as long as the mean number of trials to criterion taken by young animals. Although the range of ages of the older rabbits in this study was limited to just over 1 year, the correlation between trials to criterion and age (in months) was high ($r = .67$; $p < .05$). All but one of the older rabbits attained criterion to the tone CS within 100 trials, indicating that some transfer from the dorsolateral pontine nucleus stimulation CS occurred.

This study demonstrates that stimulation of the dorsolateral pontine nucleus mossy fibers can be an effective CS for training an eyelid response in rabbits older than 2½ years. However, acquisition is five times slower than in 3-month-old rabbits. These results indicate that the age differences in eyelid classical conditioning which have been observed in rabbits cannot be entirely attributed to age differences in the peripheral CS pathway. Changes in the nervous system occurring more centrally in the cerebellum appear to be more important for age differences in learning.

NEURONAL LEVEL ANALYSIS OF LEARNING, MEMORY, AND AGING

In the rabbit eyelid model system, the basic storage system appears to involve quite discrete cerebellar circuits, and the septohippocampal system seems to play an important modulatory role in the formation of memory. There is much evidence implicating the hippocampus in the modulation of memory storage in humans. Although the evidence does not yet exist, we would tentatively generalize from rabbit (and cat) to human regarding the cerebellar circuit essential for basic associative learning of discrete responses. As we have seen, aging has the same effects on eyelid conditioning in rabbits and humans; so it would appear that the learned eyelid response in the rabbit provides a model where both basic memory storage processes and modulatory phenomena and their interactions can be analyzed at the neuronal level in terms of the effects of aging on the formation of memories. Even in his most optimistic moments as he was running aging rats in the 1950s, James Birren would probably not have predicted that he would launch research on behavior, aging, and the nervous system to achieve this degree of progress in just three short decades.

REFERENCES

Barnes, C. A. (1979). Memory deficits associated with senescence: A neurophysiological and behavioral study in the rat. *Journal of Comparative and Physiological Psychology, 93,* 74–104.

Berger, T. W., Alger, B. E., & Thompson, R. F. (1976). Neuronal substrate of classical conditioning in the hippocampus. *Science, 192,* 483–485.

Berger, T. W., Rinaldi, P., Weisz, D. J., & Thompson, R. F. (1983). Single unit analysis of different hippocampal cell types during classical conditioning of the rabbit nictitating membrane response. *Journal of Neurophysiology, 59,* 1197–1219.

Berger, T. W., & Thompson, R. F. (1978a). Identification of pyramidal cells as the critical elements in hippocampal neuronal plasticity during learning. *Proceedings of the National Academy of Science, 75,* 1572–1576.

Berger, T. W., & Thompson, R. F. (1978b). Neuronal plasticity in the limbic system during classical conditioning of the rabbit nictitating membrane response. I. The hippocampus. *Brain Research, 145,* 323–346.

Birren, J. E. (1961). A brief history of the psychology of aging. *Gerontologist, 1,* 67-77.

Birren, J. E. (1970). Toward an experimental psychology of aging. *American Psychologist, 25,* 124–135.

Birren, J. E., & Botwinick, J. (1955). Age differences in finger, jaw, and foot reaction time to auditory stimuli. *Journal of Gerontology, 10,* 429–432.

Birren, J. E., & Wall, R. D. (1956). Age changes in conduction velocity, refractory period, number of fibers, connective tissue space, and blood vessels in sciatic nerve of rats. *Journal of Comparative Neurology, 104,* 1-16.

Birren, J. E., & Woodruff, D. S. (1983). Aging: Past and future. In D. S. Woodruff & J. E. Birren (Eds.), *Aging: Scientific perspectives and social issues* (2nd ed., pp. 1-15). Monterey, CA: Brooks/Cole, 1983.

Braun, H. W., & Geiselhart, R. (1959). Age differences in the acquisition and extinction of the conditioned eyelid response. *Journal of Experimental Psychology, 57,* 386–388.

Cohen, D. H. (1980). The functional neuroanatomy of a conditioned response. In R. F. Thompson, L. H. Hicks, & B. V. Shryrkov (Eds.), *Neural mechanisms of goal-directed behavior and learning* (pp. 283–302). New York: Academic Press.

Donegan, H. H., Lowry, R., & Thompson, R. F. (1983). Effects of lesioning cerebellar nuclei on conditioned leg-flexion responses. *Society for Neuroscience Abstracts, 9,* 331.

Drori, D., & Folman, Y. (1969). The effect of mating on the longevity of male rats. *Experimental Gerontology, 4,* 263–266.

Drori, D., & Folman, Y. (1976). Environmental effects on longevity in the male rat: Exercise, mating, castration, and restricted feeding. *Experimental Gerontology, 11,* 25–32.

Ellis, R. S. (1920). Norms for some structural changes in the human cerebellum from birth to old age. *Journal of Comparative Neurology, 32,* 1–34.

Fox, R. R. (1980). The rabbit (*Oryctolagus cuniculus*) and research on aging. *Experimental Aging Research, 6,* 235–248.

Gakkel, L. B., & Zinina, N. V. (1953). Changes of higher nerve function in people over 60 years of age. *Fiziologicheskii Zhurnal, 39,* 533–539.

Gormezano, I. (1972). Investigations of defense and reward conditioning in the rabbit. In A. H. Black & W. F. Prokasy (Eds.), *Classical conditioning II: Current research and theory* (pp. 151-181). New York: Appleton-Century-Crofts.

Gormezano, I., Schneiderman, N., Deaux, E., & Fuentes, I. (1962). Nictitating membrane: Classical conditioning and extinction in the albino rabbit. *Science, 138,* 33–34.

Graves, C. A., & Solomon, P. R. (1985). Age related disruption of trace but not delay classical conditioning of the rabbit's nictitating membrane response. *Behavioral Neuroscience, 99,* 88–96.

Haley, D. A., Lavond, D. G., & Thompson, R. F. (1983). Effects of contralateral red nuclear lesions on retention of the classically conditioned nictitating membrane/eyelid response. *Society for Neuroscience Abstracts, 9,* 643.

Harms, J. W. (1944). Altern und Somatod der Zellverbandstiere. *Zeitschrift fur Alternsforschung, 5,* 73–126.

Harrison, J., & Buchwald, J. (1983). Eyeblink conditioning deficits in the old cat. *Neurobiology of Aging, 4,* 45–51.

Hernandez, L. L., Buchanan, S. L., Powell, D. A., & Shah, N. S. (1979). A comparison of biogenic amine concentrations in discrete brain areas of "old" and "young" rabbits. *IRCS Medical Science, 7,* 356.

Hilgard, E. R., & Marquis, D. G. (1940). *Conditioning and learning.* New York: Appleton.

Ito, M. (1982). Cerebellar control of the vestibulo-occular reflex: Around the flocculus hypothesis. *Annual Review of Neuroscience, 5,* 275–296.

Jerome, E. A. (1959). Age and learning—experimental studies. In J. E. Birren (Ed.), *Handbook of aging and the individual* (pp. 655–699). Chicago: University of Chicago Press.

Kandel, E. R., & Spencer, W. A. (1968). Cellular neurophysiological approaches in the study of learning. *Physiology Review, 48,* 65–134.

Kimble, G. A., & Pennypacker, H. S. (1963). Eyelid conditioning in young and aged subjects. *Journal of Genetic Psychology, 103,* 283–289.

Landfield, P. W., & Lynch, G. (1977). Impaired monosynaptic potentiation in *in vitro* hippocampal slices from aged, memory-deficient rats. *Journal of Gerontology, 32,* 523–533.

Lavond, D. G., & Thompson, R. F. (1984). [Retest performance of young rabbits]. Unpublished raw data.

Madden IV, J., Haley, D. A., Barchas, J. D., & Thompson, R. F. (1983). Microinfusion of picrotoxin into the caudal red nucleus selectively abolishes the classically conditioned nictitating membrane/eyelid response in the rabbit. *Society for Neuroscience Abstracts, 9,* 830.

Malamud, N. (1972). Neuropathology of organic brain syndrome associated with aging. In C. Gaitz (Ed.), *Aging and the brain* (pp. 63–88). New York: Plenum.

Marwaha, J., Hoffer, B. J., & Freedman, R. (1981). Changes in noradrenergic neurotransmission in rat cerebellum during aging. *Neurobiology of Aging, 2,* 95–98.

Mauk, M. D., & Thompson, R. F. (1984). Classical conditioning using stimulation of the inferior olive as the unconditioned stimulus. *Society for Neuroscience Abstracts, 10,* 122.

McCormick, D. A., Clark, G. A., Lavond, D. G., & Thompson, R. F. (1982). Initial localization of the memory trace for a basic form of learning. *Proceedings of the National Academy of Science, 79*(8), 2731–2742.

McCormick, D. A., Lavond, D. G., & Thompson, R. F. (1982). Concomitant classical conditioning of the rabbit nictitating membrane and eyelid responses: Correlations and implications. *Physiology and Behavior, 28,* 769–775.

McCormick, D. A., Steinmetz, J. E., & Thompson, R. F. (1985). Lesions of the inferior olivary complex cause extinction of the classically conditioned eyeblink response. *Brain Research, 359,* 120–130.

McCormick, D. A., & Thompson, R. F. (1984a). Cerebellum: Essential involvement in the classically conditioned eyelid response. *Science, 223,* 296–299.

McCormick, D. A., & Thompson, R. F. (1984b). Neuronal responses of the rabbit cerebellum during acquisition and performance of a classically conditioned nictitating membrane/eyelid response. *Journal of Neuroscience, 4,* 2811–2822.

Nandy, K. (1981). Morphological changes in the cerebellar cortex of aging *Macaca nemestrina. Neurobiology of Aging,* 2, 61–64.

Norman, R. J., Buchwald, J. S., & Villablanca, J. R. (1977). Classical conditioning with auditory discrimination of the eyeblink in decerebrate cats. *Science, 196,* 551–553.

Powell, D. A., Buchanan, S. L., & Hernandez, L. L. (1981). Age related changes in classical Pavlovian conditioning in the New Zealand albino rabbits. *Experimental Aging Research, 7,* 453–465.

Powell, D. A., Buchanan, S. L., & Hernandez, L. L. (1984). Age related changes in Pavlovian conditioning: Central nervous system correlates. *Physiology and Behavior, 32,* 609–616.

Rogers, J., Silver, M. A., Shoemaker, W. J., & Bloom, F. E. (1980). Senescent changes in a neurobiological model system: Cerebellar Purkinje cell electrophysiology and correlative anatomy. *Neurobiology of Aging, 1,* 3–11.

Rogers, J., Zornetzer, S. F., & Bloom, F. E. (1981). Senescent pathology of cerebellum: Purkinje neurons and their parallel fiber afferents. *Neurobiology of Aging, 2,* 15–25.

Rogers, J., Zornetzer, S. F., Bloom, F. E., & Mervis, R. E. (1984). Senescent microstructural changes in rat cerebellum. *Brain Research, 292,* 23–32.

Scheibel, M. E., & Scheibel, A. B. (1975). Structural changes in the aging brain. In H. Brody, D. Harman, & J. M. Ordy (Eds.), *Aging* (Vol. 1, pp. 11–37). New York: Raven.

Schneiderman, N., Fuentes, I., & Gormezano, I. (1962). Acquisition and extinction of the classically conditioned eyelid response in the albino rabbit. *Science, 136,* 650–652.

Solomon, P. R., Lewis, J. L., LoTurco, J. J., Steinmetz, J. E., & Thompson, R. F. (1986). The role of the middle cerebellar peduncle in acquisition and retention of the rabbit's classically conditioned nictitating membrane response. *Bulletin of the Psychonomic Society, 24,* 75–78.

Solomon, P. R., & Moore, J. W. (1975). Latent inhibition and stimulus generalization of the classically conditioned nictitating membrane response in rabbits (*Oryctolagus cuniculus*) following dorsal hippocampal ablations. *Journal of Comparative and Physiological Psychology, 89,* 1192–1203.

Squire, L. R. (1982). The neuropsychology of human memory. *Annual Review of Neuroscience, 5,* 241.

Steinmetz, J. E., Lavond, D. G., & Thompson, R. F. (1985a). Classical conditioning of the rabbit eyelid response with mossy fiber stimulation as the conditioned stimulus. *Bulletin of the Psychonomic Society, 28,* 245–248.

Steinmetz, J. E., Lavond, D. G., & Thompson, R. F. (1985b). Classical conditioning of skeletal muscle responses with mossy fiber stimulation CS and climbing fiber stimulation US. *Society for Neuroscience Abstracts, 11,* 982.

Steinmetz, J. E., Rosen, D. J., Chapman, P. R., Lavond, D. G., & Thompson, R. F. (1986). Classical conditioning of the rabbit eyelid response with a mossy fiber stimulation CS. I. Pontine nuclei and middle cerebellar peduncle stimulation. *Behavioral Neuroscience, 100,* 871–880.

Steinmetz, J. E., Rosen, D. J., Woodruff-Pak, D. S., Lavond, D. G., & Thompson, R. F. (1986). Rapid transfer of training occurs when direct mossy fiber stimulation is used as a conditioned stimulus for classical eyelid conditioning. *Neuroscience Research, 3,* 606–617.

Thompson, R. F. (1983). Neuronal substrates of simple associative learning: Classical conditioning. *Trends in Neurosciences, 6,* 270–275.

Thompson, R. F., Berger, T. W., Cegavske, C. F., Patterson, M. M., Roemer, R. A., Teyler, T. J., & Young, R. A. (1976). A search for the engram. *American Psychologist, 31,* 209–227.

Thompson, R. F., McCormick, D. A., & Lavond, D. G. (1986). Localization of the essential memory trace system for a basic form of associative learning in the mammalian brain. In S. Hulse (Ed.), *One hundred years of psychological research in America* (pp. 125–171). Baltimore: Johns Hopkins University Press.

Thompson, R. F., & Woodruff-Pak, D. S. (1987). A model system approach to age and the neuronal bases of learning and memory. In M. W. Riley, J. D. Matarazzo, & A. Baum (Eds.), *The aging dimension* (pp. 49–76). Hillsdale, NJ: Erlbaum.

Tomlinson, B. E. (1972). Morphological brain changes in non-demented old people. In H. M. Van Praag & A. K. Kalverboer (Eds.), *Aging of the central nervous system*. New York: De Ervon F. Bohn.

Tsukahara, N. (1981). Synaptic plasticity in the mammalian central nervous system. *Annual Review of Neuroscience, 4,* 351–379.

Woodruff, D. S. (1985). Arousal, sleep, and aging. In J. E. Birren & K. W. Schaie (Eds.), *Handbook of the psychology of aging* (2nd ed., pp. 261–295). New York: Van Nostrand Reinhold.

Woodruff-Pak, D. S., Lavond, D. G., Logan, C. G., Steinmetz, J. E., & Thompson, R. F. (1985). The continuing search for a role of the cerebellar cortex in eyelid conditioning. *Society for Neuroscience Abstracts, 11,* 333.

Woodruff-Pak, D. S., Lavond, D. G., Logan, C. G., & Thompson, R. F. (1987). Classical conditioning in 3-, 30-, and 45-month-old rabbits: Behavioral learning and hippocampal unit activity. *Neurobiology of Aging, 8,* 101–108.

Woodruff-Pak, D. S., Lavond, D. G., & Thompson, R. F. (1985). Trace conditioning: Abolished by cerebellar nuclear lesions but not lateral cerebellar cortex aspirations. *Brain Research, 348,* 249–260.

Woodruff-Pak, D. S., Steinmetz, J. E., & Thompson, R. F. (1986). Classical conditioning of rabbits 2½ years old using mossy fiber stimulation as a CS. *Society for Neuroscience Abstracts, 12,* 1315.

Woodruff-Pak, D. S., & Thompson, R. F. (1985). Classical conditioning of the eyelid response in rabbits as a model system for the study of brain mechanisms of learning and memory in aging. *Experimental Aging Research, 11,* 109–122.

Woodruff-Pak, D. S., & Thompson, R. F. (1986). Delay classical conditioning of the human eyelid response with an auditory CS in 20-, 40-, and 60-year-olds. *Gerontologist, 26,* 90.

Woodruff-Pak, D. S., & Thompson, R. F. (1988a). Cerebellar correlates of classical conditioning across the life span. In P. B. Baltes, R. M. Lerner, & D. M. Featherman (Eds.), *Life-span development and behavior* (Vol. 9, pp. 1–37). Hillsdale, NJ: Erlbaum.

Woodruff-Pak, D. S., & Thompson, R. F. (1988b). Classical conditioning of the eyeblink response in the delay paradigm in adults aged 18–83 years. *Psychology and Aging, 3,* 219–229.

Woody, C. D., Yarowsky, P., Owens, J., Black-Cleworth, P., & Crow, T. (1974). Effect of lesions of coronal motor areas on acquisition of conditioned eye blink in the cat. *Journal of Neurophysiology, 37,* 385–394.

Yeo, C. H., Hardiman, M. J., & Glickstein, M. (1984). Discrete lesions of the cerebellar cortex abolish classically conditioned nictitating membrane response of the rabbit. *Behavioral Brain Research, 13,* 261–266.

9

My Perspective on Research on Aging

James E. Birren

It is appropriate and pleasant at the beginning of this personal perspective to acknowledge the fine education my colleagues and students tried to give me as I was subject to their influence. This is not an account of the limited success they may have had. This is my personal account of my perspective on the field of aging, how it evolved, something of the current content, and where it is headed.

In many respects I have been lucky, since I began my professional career as a research psychologist at the same time that professional and scientific meetings were beginning to be organized around the issues of aging. I made my first postdoctoral presentation of research at the first annual meeting of the Gerontology Society, held in New York on January 8–9, 1949 (Gerontological Society, 1948). By present standards, the meeting was a small one, as shown by the fact that the program listed the titles of only 24 papers. The printed program was only 16 pages long, and that included the abstracts and the entire list of the society's then 217 members. Of course, working in the division of gerontology of the National Heart Institute, my colleagues and I were encouraged to join the Gerontological Society and to participate in scientific meetings. I also participated in the first meeting of the Division of Maturity and Old Age (then its name) of the American Psychological Association, in Detroit in 1947, and the first National Conference on Aging, sponsored by the federal government in 1950 (U.S. Government, 1951).

The first National Conference on Aging, the forerunner of the White House Conferences on Aging, was sponsored and organized by the Federal Security Agency (which became the Department of Health, Education and Welfare and still later the Department of Health and Human Services). The conference began on August 13, 1950, and had 816 participants. It was pointed out in the conference report that "the only government supported research program in the field of gerontology, as such, is that of the Public Health Service, National Heart Institute, in the division of gerontology" (U.S. Government, 1951). The report made a further, although somewhat contradictory, statement:

> Some research on the problems of aging is also being conducted by the Federal Government. Among 26 government agencies which were presumed to have problems within the field of gerontology, 15 reported research activities in this area. It appears that, for the most part, problems in aging are investigated in these agencies only sporadically and in conjunction with other studies. (U.S. Government, 1951, p. 246)

Just as my career was starting, other activities were beginning to be organized quickly. For example, the International Conference on Gerontology was organized in Liège, Belgium, July 10–12, 1950, with plans leading to the international conference held in St. Louis, Missouri, in 1951, which I attended as a still relatively young career researcher.

It should be pointed out that the American Geriatric Society was founded in 1942 and the Gerontological Society in 1944, but these were war years and there was no opportunity to hold annual meetings or to encourage research and scholarly exchange on the issues of aging, which were priorities remote from the war effort.

THE GROWING REALIZATION OF THE NEED FOR RESEARCH ON AGING

These early years had a strong impact on me insofar as I became increasingly convinced of the necessity to encourage research on the issues of aging. Since that time, research on aging has grown exponentially in response to the increasing number of older adults in our society, the lower birth rate, and the resultant need to make better use of older adults' potential. These trends have continued and escalated, and the need for research on aging continues to grow.

During the formative years of my career, and the start of professional societies, a second conviction which I developed was that the problems of aging are in most cases not like the lock-and-key problems of many areas (e.g., poliomyelitis and immunization), in which there is one single solution. They become transformed over time and as our standards and expectations rise, even as the topics seem to remain the same. The agenda for the first National Conference on Aging is remarkably like the agenda topics for the White House Conferences on Aging that were subsequently held each 10 years. There were 11 topics identified for consideration in 1950: (1) population changes and economic implications; (2) income maintenance; (3) employment, employability, and rehabilitation; (4) health maintenance and rehabilitation; (5) education for an aging population; (6) family life, living arrangements, and housing; (7) creative and recreational activities; (8) religious programs and services; (9) professional personnel; (10) aging research; and (11) community organization. These themes are found in subsequent White House Conference reports as well as in the published literature of the field.

POLITICIZATION

It was perhaps inevitable that gerontology would become increasingly politicized as interest rose in the older voters and as the older market increased. The predicted rise in the number and proportion of older persons in our society did indeed become apparent, and along with it debates arose about the responsibilities of the federal government, of the states, and of the private sector for the well-being of older people.

By the time of the 1981 White House Conference on Aging, political interest boiled over. Partly this was due to the fact that the White House Conference (WHC) was initially planned under a Democratic administration and then modified and carried out under a Republican administration. The former, with its traditions going back to the Social Security and New Deal of the 1930s, tended to emphasize the role of the public sector, whereas the new Republican administration was emphasizing private sector initiatives and minimizing the role of the federal government.

Gone by 1981 was the relaxed friendliness of earlier days when scholarly activities and research occupied a back bench on the national scene.

The 1981 WHC had security personnel with walkie-talkies, and the tone of the discussions changed dramatically. Long-standing colleagues stopped speaking to one another, while rumors of political control of the conference were rife. The original chair of the organizing committee was a former member of Congress from California, Jerry Waldie. At the time of the WHC, it was rumored that he had been denied admission to the main forum and had had to slip in via a fire escape to attend. When a former colleague of mine was asked to identify the political affiliation of potential chairpersons for WHC committees before their appointments, I realized that my colleague had been put into a questionable role by higher authorities.

Although I was the chair of the technical committee that prepared the background statement on research, I was not given discussion or voting privileges on research at the conference. The context of aging had changed between the first meeting of the Gerontological Society in 1948 and the WHC of 1981.

AGING AND THE MARKET PLACE

The issues of an aging population and aging individuals are now mainstream matters, and it is not likely that we will see much reduction in political concerns of elections and legislation or self-interest marketing in the field. To me this requires a rising ethical sensitivity in the academic community, in researchers, and in professional personnel about their personal interests, corporate interests, and the public good. The growth of ethical concerns in the field of aging is not restricted to matters of terminal care of the dying person. There are broader areas of concern, such as the equitable distribution of services by age, sex, and ethnicity.

We will probably have to be more aware of the implications of the support by for-profit institutions of researchers in nonprofit universities and institutes. Big marketplace dollars can attract questionable enterprises. Scholars and professionals with very little experience in matters of professional ethics may be drawn into research activities that are ethically questionable. In brief, I think that ethical issues in aging research and practice increasingly will be discussed and questioned.

POLICY ISSUES

Echoing ethical issues arising out of aging populations are matters of health prevention that touch upon marketplace economic interests. I was amazed to learn in China in September, 1986 that cigarette smoking was not discouraged, although it is recognized that smoking promotes lung cancer as well as cardiovascular disease. The policy decision in China was based upon the fact that the capital produced by cigarettes was wanted for investment in other areas of society. The conflict was in the support of public priorities requiring capital rather than in the reduction of smoking for individual health improvement. Health savings gained through a no-smoking policy were not regarded as spendable.

In American society there are tensions between those interested in public health and those interested in economic gains from tobacco growing (subsidized) and the manufacturing and sales of cigarettes. During my career these tensions have continued to exist, and in my opinion, public policy has often trailed expert knowledge by a considerable gap. Alcohol use and alcohol abuse is another area where private-sector interests and public health policies intersect. While many issues of food habits and toxicity are related to their effects on the young, there are also accumulative effects of small exposures over a lifetime that in late middle age or old age contribute to morbidity and mortality. Such long-term accumulation or "sleeper" effects also evoke policy concerns about public versus private interests.

The fair participation in the gross national product of members of our society by age, sex, and ethnicity is a topic of continuing concern. Matters of the size of Social Security payments, pension benefits, and the extent of public support of health care will continue to grow in significance, and public debates should reflect better information. In this process, universities increasingly will have to develop approaches, methods, and experience with policy analysis as a background for public debates about age-related policy decisions by state and federal legislatures. The one-at-a-time decision style about issues affecting an aging population may have been appropriate several decades ago, but now our legislative bodies, as well as the private sector, need more comprehensive approaches to policy analysis.

My personal perspective also suggests that the time span of our policy analysis is too short. We have a national rhythm determined by the 4-year presidential elections. Thus, our policy time span tends to be limited to 4 years and tends to be excessively driven by political philosophies and election prospects rather than by rational analyses of issues. Helpful in my opinion would be several policy study institutes in which major attention can be devoted to issues of aging, to inform policy on the basis of longer historical periods as well as future trends, so that our future policies will have not only the advantage of historical analyses but also more carefully considered explorations of the decades ahead of us.

GERONTOLOGY AND THE UNIVERSITY

During the 40 years of my career I have come to recognize the considerable length of time it takes to build up human organizations that have the human resources of trained personnel who will undertake research in the field of aging. Individual careers are short relative to the long life spans of institutions. Thus, I have been at my present university, the University of Southern California, for 22 of its 107 years of existence. This is a little over 20% of the institution's time, and it is only during that period that it has had an organized research and educational activity in the field of aging.

The qualities of institution building appear to draw out different elements than does career building. More cooperation than competition is called for in building institutions. Also, young, highly motivated, and well-trained investigators are encouraged to undertake their careers in gerontology as well as in many other fields, but they need short-term payoffs. Promotion to tenure in a university is usually between 3 and 7 years after initial employment. During this time, the individual is supposed to make substantial contributions and show evidence of productivity worthy of career employment. However, not all subject matters, including some aspects of gerontology, fit the career need to produce rapidly and establish an early reputation. Many young investigators may look at the career prospects and weigh them in terms of the likelihood that they will get a scientific and personal payoff in a short period of time relative to the long-term interests of institutions in building traditions of support for gerontological research.

In recent years, the emphasis on fiscal policies in universities has created a climate within which the researcher must be not only intellectually productive early in his or her career but economically productive for the institution as well. At USC, academic units were designated in 1982 as revenue centers, with short-term budget fluctuations having a consequence for the units. Should the pressure for a revenue center concept of academic units continue, it is likely that it will reinforce the need for federal government research institutions, the orientation, policies, and traditions of which can support research over many decades, buffeted from transient fiscal oscillations or marketplace concerns.

INSTITUTIONAL DEVELOPMENT

It is apparent that there are some special requirements in research on aging that do not always intersect or mesh well with early individual career development or institutional development. In the pursuit of short-range goals, institutions may elect to initiate research on aging. But the requirements for sample development and data collection, as well as the maturing development of investigators, require more time than is possible to compress into an approach of "Well, let's try it for three years and see how it goes."

It is worth noting that the federal government's Gerontology Research Center was founded in 1941 and the publication of the Baltimore Longitudinal Study of Aging was in November 1984 (NIH, 1984). Thus, there were 43 years between the idea and initial planning and the eventual realization in terms of a significant contribution to the understanding of human aging, a slow rate of return. It is perhaps not surprising, therefore, that major organized research efforts on aging in America were sponsored and carried out in government laboratories. Although extramural branches of the National Institutes of Health have supported many individual investigators in universities and other groups, for a long time foundation and private sector support was at a minimum in the field of aging. An early exception was the Josiah Macy, Jr. Foundation, which established a conference program on problems of aging in 1937. The collegial interactions fostered by that foundation led to the organization of the Gerontological Society, as well as the activities of E. V. Cowdry, which led to the publication of the first organized volume on issues of aging, *Problems of Ageing* (Cowdry, 1939). It was from this foundation that the field took the stance of addressing issues of aging from a multidisciplinary perspective.

Partly, it is the multifaceted nature of aging that confounds the departmentalized approach of many universities and colleges which are organized around teaching obligations and narrow professional and disciplinary interests. As a result, research and scholarly activities on aging often are fragmented in academic institutions according to the self-interests of investigators and departments. There is little doubt that almost all of the disciplines and professions have a contribution to make to the understanding of aging. These special fields have to be encouraged in their efforts, along with the more integrated efforts implied by the organization of institutes and centers that transcend disciplinary boundaries.

My perspective comes from experience as a career scientist in several of the National Institutes of Health—the National Heart Institute, the

National Institute of Mental Health, and the National Institute of Child Health and Human Development. In addition, I have had the opportunity of carrying out research and teaching at the University of Chicago and at the University of Southern California. Strengths and weaknesses of both government institutions and universities prompt me to express the opinion that both are very important in promoting understanding of aging through research. The creation of the National Institutes of Health has been one of the great contributions to scientific research in America, and I have great respect for the wisdom and planning that went into their development. Given the long history of universities, there is less need to single them out here for comment about their role in knowledge generation and education. As discussed in the earlier section on policy issues, however, both federal research programs and universities have to be protected from the short-term political and marketplace interests that have come to be associated with aging.

Issues of human aging will be with humankind throughout its existence. In this sense, we are only at the early stage of institutionalized research efforts to understand the many processes of aging. How humankind grows up and grows old is a confluence of forces that include the evolution of species characteristics, individual heredity, the physical and social environments in which we live, and those internal balances and dispositions that lead to our choices of life-styles and behavior. For me, the appropriate way to express this confluence is to describe human aging as a broad issue in ecology.

THE ECOLOGY OF AGING

Defining human aging as an ecological phenomenon implies to me that there will be shifting influences on aging as we pass through different historical eras. For example, at the beginning of this century, sex differences in mortality and life expectancy were trivial. Now, greater female gain in life expectancy is creating social consequences as it hovers between 7 and 8 years of longer life expectancy for females than for males. Whereas this sex difference was a trivial variable earlier, today it is a major one. By contrast, the pervasive acute and long-term consequences of infectious diseases that occurred in childhood during the early part of this century are now disappearing. Thus, if we take the length of life and the occurrence of disease with age as something to be explained, there necessarily has to be a shift in the relative importance of different factors with historical period.

Many environmental circumstances can modulate the expression of both species and unique individual heredity. Environmental circumstances may also reach a peak and decline, such as will be seen regarding the consequences of smoking and alcohol use if public health measures are successful. The ecology of human aging implies that there is a common thread running through questions we are asking. For me the basic query is, how are the phenomena of aging organized?

Better public information has produced more intelligent consumers, who recognize that there are upper limits to the benefits to be derived from exercise, better nutrition, and the avoidance of smoking and alcohol and drug abuse. Nevertheless, the public is convinced that the length of our lives, the results of age-associated illness, and the quality of social and psychological life potentially can be improved through the application of knowledge gained through scientific research. Much of what kills us appears to be under our control, although what governs our rate of aging remains obscure.

There has been a trend toward recognizing that human aging is a multifactorial phenomenon. Less tolerated today is the uninformed enthusiast who believes he or she has *the* single variable answer to the question of what is human aging. Another trend that appears to characterize the field is a movement toward accepting social and behavioral variables such as occupation, education, and sedentary life-styles as significant contributions to well-being over the life span. In retrospect it appears to me that we researchers were simplistic in our earlier views that one or two variables were dominant. Also evident in the early views of many of my colleagues were presuppositions about what was significant. One of these suppositions was that biological factors always take precedence over psychological and social factors. Partly, this issue is resolved by consideration of what we are taking as the major dependent variable in our explanations. For the dependent variable, I would consider aging in relation to the World Health Organization (WHO) definition of health, which is total physical, psychological, and social well-being.

A shift in scientific thought, however, had to occur before we could explore health and aging issues as a matter of ecology. For one thing, we had to come to accept the idea that social factors, as in the case of bereavement or the psychological stress effects of job loss, retirement, and other factors, could be causal in initiating adverse physiological change. Certainly, today stress is a commonly accepted variable in influencing adult health. Although some scientists might not appreciate the significance of social and psychological variables as important causes of outcomes in human aging, the present state of research knowledge would support acceptance of the idea that psychological and social factors can be the antecedents as well as the consequences of health changes in the later years. We should now talk about the pathways to various outcomes, implicitly accepting the idea that there are many influences.

DATA AND THEORY

My characterization of the field of aging, after having been associated with the publication of several handbooks on aging, leads me to regard it as "a land of many islands of data with few bridges between them" (Binstock & Shanas, 1985; Birren & Schaie, 1985; Finch & Schneider, 1985). This implies that we are in a phase of being data-rich and theory-poor. An investigator's first research on aging usually begins with a microtheory derived from a subpart of the discipline in which the investigator was trained. It is perhaps to be expected that the narrow view and the microtheory characterize our present phase. A practical issue is one of the investigator's making headway within a discipline to establish the benchmarks for career growth, which may not be in concert with the need to understand complex, naturally occurring problems.

Of course, one reason for the data-rich, theory-poor state of research on aging is the inherent complexity of the subject matter and a past lack of appropriate methods for study. Today one begins to hear more talk about causal models and pathway considerations. In step with the growth of sophisticated data analysis methods, design of research, and availability of samples, one may expect that there will be more attempts at an integrative theory in aging.

In the past there does not appear to have been a high level of "additivity" of information in the psychological and social sciences. That is, convergence of evidence resulting in agreed-upon points has been slower in these fields than it has been in the biological sciences. Perhaps in the biological sciences there is greater agreement or consensus about the dependent variable—that is, what is being explained—than in the social and behavioral fields. In turn, this may be reflected in the lack in the social sciences of definitions of aging as a technical term in discussions.

In explaining some of the attributes of aging, we might use the analogy of explaining the functions of a computer. In explaining the functions of a computer, there are two levels of description. One is the hardware and its principles, and the other is the software and its principles. Software and hardware languages are clearly different. In a similar way, the organization of the nervous system that occurs through learning is more akin to a software program, whereas the explanation of bodily regulation by the hypothalamus is more readily explained in terms of hardware principles. In fact, for the computer and the human brain, both experientially organized (software) and structurally organized (hardware) principles and explanations are entirely appropriate and are called for.

In addition to the "hardware" and "software" issues, there is the zone of interaction in which the human brain's experientially organized control of behavior and biological control mechanisms interact. For example, contemporary discussions about the role of human experience and stress in the functioning of the immune system with age or the role of experience and behavior in the genesis of cardiovascular disease illustrate hardware–software or biobehavioral interactions with age.

As yet, we do not have an adequate scientific language or theory to address the interactions of the experientially organized functions with the biologically organized ones. Attempts to do so in the past were sometimes noted by the use of the term *psychosomatic*. These references were usually weighted in the direction of psyche influencing soma rather than the recognition of the fact that soma and psyche may interact and that the state of the organism is a natural product of the hardware and the software, the biologically and experientially organized portions of the nervous system. There is probably no condition of human life, its length, morbidity, disability, or the contentment that an individual derives, that is not modulated by the nervous system and that therefore is not in part experientially organized.

LEARNING AND PRACTICE

We are nowhere near understanding the full significance of the consequences of learning and practice in relation to age. It has been difficult to study the lifelong consequences of practice or disuse since there have been few opportunities to follow individuals over long periods of time. However, the Baltimore Longitudinal Study (Shock et al., 1984) as well as others, such as the Berkeley Growth Studies (Jones, 1958), offers useful information about the effects of disuse and learning on human capabilities. Among the capabilities that have been discussed in such psychological literature is that of intelligence and how it may change over the course of life. Psychologists have perhaps put more effort into the measurement of intelligence than of any other human quality. Information in this volume points to many of the properties of intellectual ability and how they may change with age and under environmental circumstances that encourage disuse or use. However, we are still a long way from understanding in detail how intellect, which enables us to adapt to a changing environment, is related to our probabilities of living long and free of disease. Here I would like to see intelligence measured in the context of the well-being of the organism as implied by the WHO definition of health, rather than measured by the older criterion of the capacity to learn by itself.

HEALTH, BEHAVIOR, AND AGING

Psychologists have been reluctant until recently to place their measurements in the context of health, as though health were a foreign domain without behavioral significance. However, they are realizing more and more that health and behavior are interlocked, and the last decade has seen the publication of many books that illustrate the coupling between health and behavioral variables. The field of health psychology is burgeoning and needs co-opting by the concerns of aging.

The well-being of the organism is perhaps best expressed by the capacity of the organism to adapt and change in relation to environmental demands. With the growth of research on dementing disorders, neuropsychology has risen to the fore in the need to identify the locus and nature of disturbed processes in the central nervous system. In a sense, this is research on intelligence revisited, in which there is a reexamination of the components of the intellect in relation to health, whereas in the early context the measurement of intellectual factors focused on the prediction of school achievement or the likelihood of learning.

The growth of neuroscience and the neuropsychology of aging is one excellent opportunity to integrate both the biologically programmed portion of the nervous system and the experientially programmed part. The power of our generalizations about aging will increase with the number of contexts in which the phenomena have been observed.

LOCALIZATION OF CAUSAL MECHANISMS

Another trend in neuroscience is in the localization of mechanisms, that is, identifying the steps in the pathways through which different processes are linked. There remains an active search for the anatomical, biochemical, and functional attributes of learning. There seems little doubt that results of fundamental studies of the mechanisms of learning and extinction will contribute to our understanding of aging. The prospects for this understanding are influenced by the strength of our desire to explain and integrate data rather than to prove the superiority of one set of variables over another.

It has been over 100 years since Francis Galton (1885) gathered data on the behavioral characteristics of a wide age range of people in a British health exposition in London. Still awaiting explanation by the inquiring mind are the reported associations of age, speed of response, and vital capacity. This presupposes, however, that there is a desire to explain such correlates and that a cooperative attitude exists to link broad domains of data.

THE ORIENTATION OF GERONTOLOGISTS

The foregoing comments are in part a call for collaborative efforts between investigators trying to explain how aging is organized. In addition, they represent a call for greater collaboration between scientists of different backgrounds. It seems to me, after 40 years of experience in gerontology, that the competitive edge of explanation has been fostered to the detriment of cooperative effort to achieve explanation. Aging embraces vast domains of complex variables, and its explanation and modeling require the cooperation of different disciplines.

Perhaps our spawning grounds of trained investigators in universities have especially stressed competition of an intellectual sort, where individual achievement is rewarded. The competitive versus the cooperative climate of the intellectual environment is relevant to the thought that investigators become more expert in analysis and dissecting phenomena than in synthesis, or putting them together, when the latter effort requires collaboration of other scientists. The main point I am musing about is the research and intellectual atmosphere of our training institutions. Do we excessively train for competition, and can we train simultaneously for individual excellence as well as for cooperation on complex projects of human concern? Certainly the subject matter of aging is complex, and we need to mobilize multidisciplinary cooperation in its study.

THE FUTURE

There seems little doubt that research on aging in the foreseeable future will remain a focal point within the biological, behavioral, and social sciences. Society will expect more information from the intellectual community about ways to promote lifelong biological, psychological, and social well-being. This is an exciting scientific period, when partial answers are beginning to be seen about the conditions which limit human productivity and which may lead to excessive mortality, morbidity, and disability over the adult years. There are many steps of a scientific character that must be taken before we can with certainty say what the pathways to productivity and individual well-being are in the later years.

Surrounding us are better-educated and informed adults who want to learn from the scientific community what efforts they can exert to maximize their productivity and healthy life and minimize the consequences of adverse morbidity, problems of mental health, and physical disability. Increasingly, society has become the patron of the scientist interested in aging. The study of aging is no longer a sleepy backwater of scientific activity, as it was 50 years ago. It is a field of great consequences and great challenges for the inquiring mind and progressive institution. I have been lucky, and I am grateful to have had the opportunity to have my career shaped during these early years of the emergence of gerontology. I am grateful for the stimulation and guidance of the many senior pioneer members of the field who have served as my academic mentors and colleagues. I am also grateful for the stimulation provided later in my career from participating in the intellectual growth of graduate students who elected aging as their field of study. The former students and colleagues who contributed to this volume show through their works that there are important threads in gerontology that are being pursued with sophistication and vigor. In my opinion, the field has moved ahead dramatically and will achieve more in the care of such persons.

REFERENCES

Binstock, R. H., & Shanas, E. (Eds.). (1985). *Handbook of aging and the social sciences*. New York: Van Nostrand Reinhold.

Birren, J. E., & Schaie, K. W. (Eds.). (1985). *Handbook of the psychology of aging*. New York: Van Nostrand Reinhold.

Cowdry, E. V. (Ed.). (1939). *Problems of ageing*. Baltimore: Williams & Wilkins.

Finch, C., & Schneider, E. L. (Eds.). (1985). *Handbook of the biology of aging*. New York: Van Nostrand Reinhold.

Galton, F. (1985). On the anthropometric laboratory at the late International Health Exposition. *Journal of the Anthropological Institute, 14*, 205–218.

Gerontological Society. (1948). Program of the annual meeting. *Journal of Gerontology, 3*(Suppl. 4), 1–16.
Jones, H. E. (1958). Problems of method in longitudinal research. *Vita Humana, 1,* 93–99.
National Institutes of Health. (1984). *Normal human aging.* Washington, DC: U.S. Government Printing Office. (Publication No. 84-2450)
Shock, N. W., Greulick, R. C., Andrus, R., Arenberg, D., Costa, P. T., Jr., Lakatta, E. G., & Tobin, J. D. (Eds.). (1984). *Normal human aging: The Baltimore Longitudinal Study of Aging.* Washington, DC: U.S. Government Printing Office. (NIH 84-2450)
U.S. Government. (1951). *Man and his years. An account of the first National Conference on Aging.* Sponsored by the Federal Security Agency. Raleigh, NC: Health Publications Institute.

Index